PIEZOELECTRIC CERAMICS

Principles and Applications

Second Edition

APC
International, Ltd.

Piezoelectric Ceramics: Principles and Applications

Piezoelectric Ceramics from APC International, Ltd.

APC INTERNATIONAL, LTD., formerly American Piezo Ceramics, Inc. (Est. 1986), is a leading supplier and distributor of piezoelectric ceramics and piezo devices to domestic and international companies.

Markets Served

APC International's piezoelectric powders, ceramics, and devices are used in a wide range of applications including:

- Ultrasonic medical transducers
- Ultrasonic medical transducers
- Ultrasonic cleaning components
- Ultrasonic welding components
- NDT components
- Flow control products
- Level control products
- Accelerometers
- Proximity Sensors

APC International's products support a wide range of industries including:

- Aerospace
- Automotive
- Consumer products
- Defense
- Engineering & design
- Industrial & manufacturing
- Medical devices
- Research & development

Our Products

APC International's standard products include:

- Piezoelectric ceramics in standard and custom shapes
- Piezoelectric powders
- Stack actuators
- Air transducers
- Fluid atomizers / nebulizers
- Piezo buzzers
- Disc benders / unimorphs
- Stripe™ actuators
- Piezo Ignitors

Our Manufacturing Capabilities

APC International specializes in meeting its customer's precise custom requirements for piezoelectric ceramics and piezo devices.

APC International manufactures piezoelectric discs, rings, plates, cylinders, and custom shapes from our specially formulated soft and hard-body lead-zirconate-titanate (PZT) compositions. APC International's processing capabilities include: machining, pressing, firing, electroding, poling and testing piezoelectric ceramics and piezo devices.

APC International will design and build piezo devices specific to customer applications, perform custom electroding, match ceramic parts in pairs or groups, supply soldered leads, tighten standard product tolerances, and provide electrical testing reports.

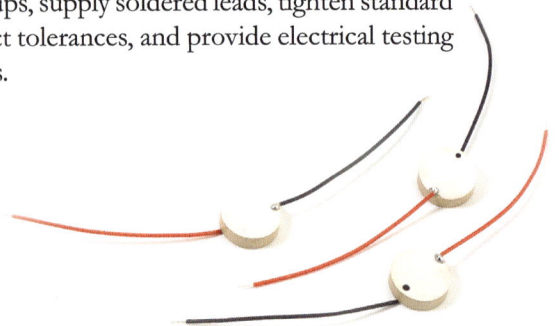

Customer Philosophy

At APC International we focus on meeting the needs of our customers. We understand that in our industry achieving this goal often requires ongoing discussions with our customers. Having an open dialogue with our customers ensures that our piezoelectric ceramics are manufactured with appropriate specifications to achieve our customer's desired results and increase the overall reliability of our customer's final product. Using a consultative sales approach, APC International strives to deliver high quality products, on time, and at a competitive price.

Worldwide Representation

APC International has sales representatives in the United States and in many countries around the world. Our representatives take the time to understand our customers' needs and work with our customers to select the most appropriate piezoelectric ceramic or piezo device. If you need a piezoelectric ceramic element or an item that incorporates a piezoelectric ceramic element please contact one of our representatives to discuss your project.

Ian R. Henderson
President, APC International, Ltd.
www.americanpiezo.com

Your Partner From Design to High Volume Production

APC International's engineering and design team of material scientists and engineers has years of experience in piezoelectric ceramic design and applications. In Pennsylvania, we have state-of-the-art manufacturing facilities and a production staff with an average of ten years of experience at the company. In addition to our manufacturing facilities in Pennsylvania, APC International has established partnerships with other companies around the world. Drawing on these resources, APC International can serve as its customer's single source supplier of piezoelectric ceramics and piezo products from the initial design stage through the high volume production stage of a product's lifecycle.

APC
International, Ltd.

Contents

Actuators

Transducers

Miscellaneous

ISBN 0-9718744-0-9

introduction

When subjected to a mechanical force certain crystalline minerals become electrically polarized. Tension and compression generate voltages of opposite polarity, in proportion to the applied force. The converse of this relationship also is true: a voltage-generating crystal exposed to an electric field lengthens or shortens according to the polarity of the field, and in proportion to the strength of the field. These behaviors are the *piezoelectric effect* and the *inverse piezoelectric effect*, respectively. Piezoelectric materials have been adapted to an impressive diversity of sensing and action applications, including generation of sonic and ultrasonic signals. For many of these applications there are no practical alternatives.

Metal oxide-based piezoelectric ceramics and other man-made materials have enabled designers to employ the piezoelectric effect and the inverse piezoelectric effect in applications for which natural materials are unsuitable. The composition, shape, and dimensions of a piezoelectric ceramic element can be tailored to meet demanding requirements. Further, these materials are physically strong and chemically inert, and they are relatively inexpensive to manufacture.

A ceramic element of suitable composition is made piezoelectric by exposing the element to a strong direct current electric field. When the electric field is removed the element is permanently polarized *(poled)* and elongated in the direction of the field. Subsequently, compression along the direction of polarization, or tension perpendicular to the direction of polarization, generates voltage of the same polarity as the polarizing voltage. Tension along the direction of polarization, or compression perpendicular to the direction of polarization, generates voltage with polarity opposite that of the polarizing voltage. This conversion of mechanical energy into electrical energy—*generator action*—is used in fuel-igniting devices, solid state batteries, and other products.

If a voltage of the same polarity as the polarizing voltage is applied to a ceramic element, in the direction of the polarizing voltage, the element will lengthen and its diameter will narrow. If a voltage of polarity opposite that of the polarizing voltage is applied, the element will become shorter and broader. If an alternating voltage is applied, the element will lengthen and shorten cyclically, at the frequency of the applied voltage. This is *motor action* – electrical energy is converted into mechanical energy. The principle is adapted to piezoelectric motors, sound or ultrasound generating devices, and many other products.

In a continuation of the technological evolution through which ceramic materials supplanted naturally piezoelectric materials, an expanding variety of next-generation piezoelectric materials is being developed for acoustical, optical, medical, wireless communication, and other applications. Piezoelectric elements fabricated from *single crystals* of lithium niobate, synthetic quartz, or other materials can exhibit significantly superior piezoelectric properties, relative to polycrystalline elements. Relative insensitivity to temperature, very high electrical energy / mechanical energy converting factors, and other attributes make *relaxor materials* very attractive for a variety of applications.

Introduction

The last decades of the nineteenth century saw several discoveries that seemed like curiosities at the time, but which now are regarded as landmarks in the evolution of modern technology. In 1880, Jacques and Pierre Curie discovered an unusual characteristic in certain crystalline minerals: when they subjected these crystals to a mechanical force, the crystals became electrically polarized. Tension and compression generated voltages of opposite polarity, and the voltage was proportional to the applied force. Soon after this discovery, it was determined that, conversely, when such crystals were exposed to an electric field they experienced an elastic strain, and lengthened or shortened according to the polarity of the field, and in proportion to the strength of the field. These behaviors were labeled the piezoelectric effect and the inverse piezoelectric effect, respectively, from the Greek word *piezein*, meaning to press or squeeze. Although these materials rarely contain iron, they often are called ferroelectric materials because their electrical behavior is analogous to the magnetic behavior of ferromagnetic materials.

The magnitudes of piezoelectric voltages, movements, or forces are small, and often require amplification (a disk of piezoelectric ceramic will increase or decrease in thickness by only a fraction of a millimeter, for example), yet the properties of piezoelectric materials have been harnessed for an impressive range of applications. The piezoelectric effect is used in sensing applications, such as in force or displacement sensors. The inverse piezoelectric effect is used in actuation applications, such as in motors and devices that precisely control positioning, and in generating sonic and ultrasonic signals.

For many years natural crystals such as quartz and tourmaline were the exclusive source of piezoelectric capabilities, and many types of products were developed with these materials. In recent decades, however, and especially since the mid 1960s, man-made materials—piezoelectric ceramics prepared from metallic oxides—have replaced natural materials in many applications, and have enabled designers to employ the piezoelectric effect and the inverse piezoelectric effect in many new applications.

A traditional piezoelectric ceramic has a perovskite crystal structure, each unit of which consists of a small, tetravalent metal ion, usually titanium or zirconium, in a lattice of large, divalent metal ions, usually lead or barium, and O^{2-} ions (Figure 1.1). Under conditions that confer a tetragonal or rhombohedral symmetry on the crystals, each crystal has a dipole moment (Figure 1.1b). The initial discovery that metallic oxides could be made piezoelectric was made with barium titanate ($BaTiO_3$) materials, but other ceramics also exhibited these properties. Subsequently, when

Figure 1.1 Crystal structure of a traditional piezoelectric ceramic

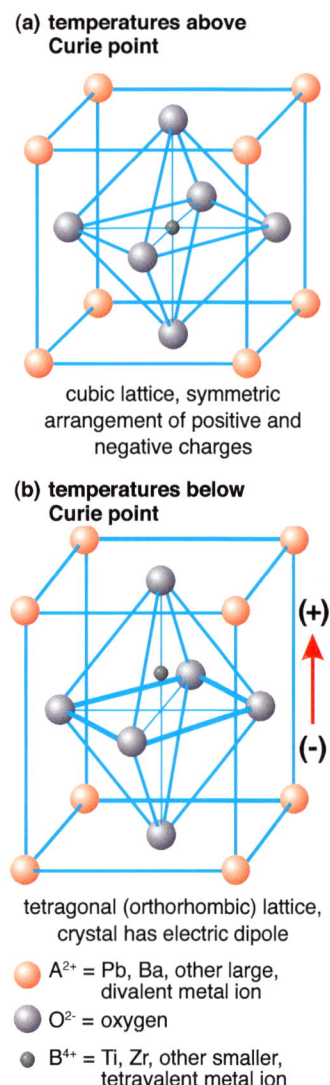

(a) temperatures above Curie point

cubic lattice, symmetric arrangement of positive and negative charges

(b) temperatures below Curie point

(+)

(-)

tetragonal (orthorhombic) lattice, crystal has electric dipole

A^{2+} = Pb, Ba, other large, divalent metal ion

O^{2-} = oxygen

B^{4+} = Ti, Zr, other smaller, tetravalent metal ion

Figure 1.2 Polarizing (poling) a piezoelectric ceramic*

| (a) random orientation of polar domains prior to polarization | (b) polarization in DC electric field | (c) remanent polarization after electric field removed |

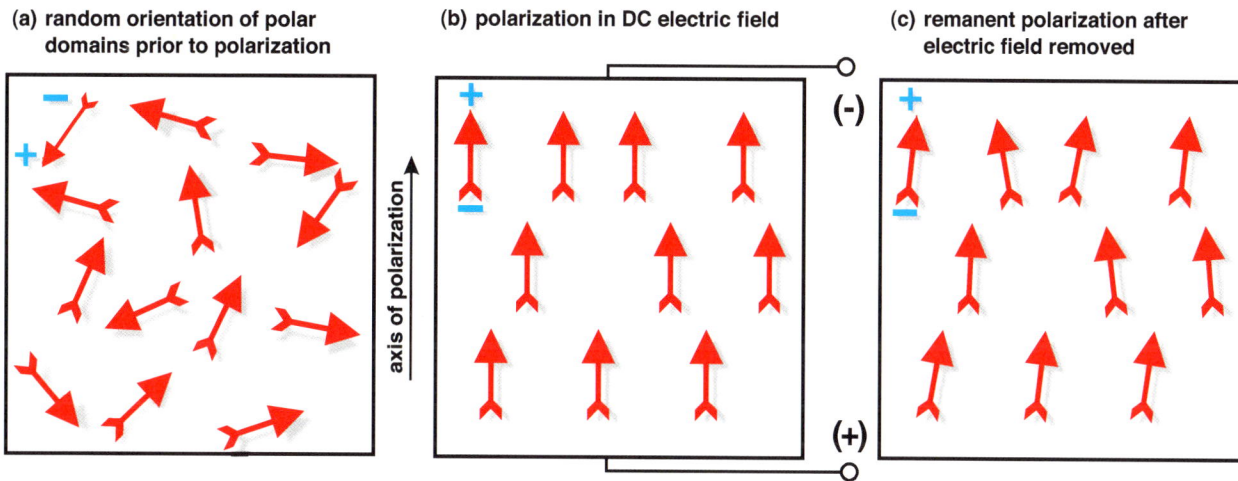

investigators determined that lead-zirconate-titanate materials exhibited greater sensitivity and higher operating temperatures, relative to $BaTiO_3$ materials, "PZT" materials – $Pb[Zr_{(x)}Ti_{(1-x)}]O_3$ – replaced $BaTiO_3$ materials in many applications, and have become the most widely used piezoelectric ceramics. Other ceramics are used for specific purposes. For example, lead titanate has characteristics that make it desirable for use in hydrophones, medical diagnostic equipment, and automobile engine knock sensors (1), but because it is difficult to polarize, lead titanate is not more widely used. Similarly, ceramics prepared from lead metaniobate are used in apparatus for nondestructive testing of materials, in medical diagnostic equipment, and in hydrophones, but manufacturers must contend with high porosity and poor mechanical strength when using these materials. Piezoelectric ceramics supplied by APC International, Ltd. are manufactured from highest purity precursors, with properties optimized for specific applications by adjusting the zirconia:titania ratio, and/or by including secondary materials.

Piezoelectric ceramics can be hundreds of times more sensitive to electrical or mechanical input than natural materials**, and the composi-

tion, shape, and dimensions of a ceramic can be tailored to meet the requirements of a specific purpose. Piezoelectric ceramics are physically strong, chemically inert, and immune to humidity or other atmospheric influences, and they can be manufactured relatively inexpensively.

To prepare a piezoelectric ceramic, fine powders of the component metal oxides are mixed in specific proportions, then heated (calcined) to form a uniform powder. The powder is mixed with an organic binder and is pressed, calendered, or molded into structural elements having the desired shape (disks, rods, plates, etc.). The "green" ceramic shapes are fired according to a specific time and temperature program, during which the powder particles sinter and the material attains a dense crystalline structure. The shapes are cooled, then further shaped or trimmed, if appropriate, and electrodes are applied to the appropriate surfaces.

A fired ceramic element is a semi-organized mass of fine crystallites (ceramic grains). A typical ceramic sample contains 10^9 to 10^{12} grains per cm^3. Above a critical temperature, the Curie point, each perovskite crystal in each grain in the element exhibits a simple cubic symmetry with no dipole moment (Figure 1.1a). At temperatures

* Simplified for clarity. Polarization does not eliminate the random distribution of grains in the ceramic. A small percentage of grains will be constrained from aligning in the direction of polarization. Changes in orientation in Figure 1.2c are exaggerated for easier visualization.

** For comparison of piezoelectric characteristics of piezoelectric ceramics and quartz crystals see Table 1.6.

Figure 1.3 Effects of electric field (E) on polarization (P) and corresponding elongation / contraction of a ceramic element

(a) hysteresis curve for polarization

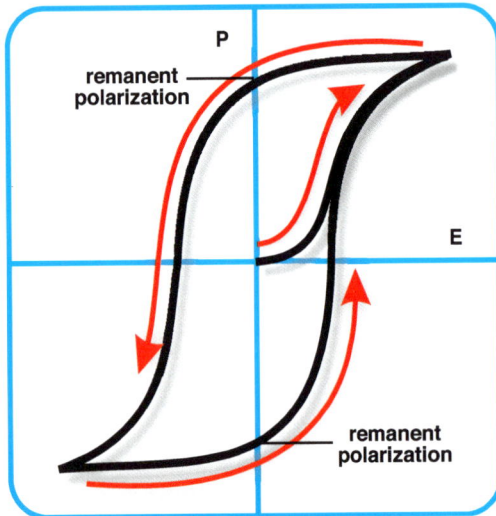

(b) relative increase/decrease in dimension (strain, S) in direction of polarization

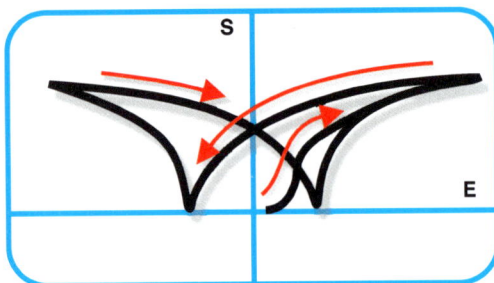

below the Curie point, however, each crystal exhibits tetragonal or rhombohedral symmetry, depending on the composition of the material, and its structure carries a dipole moment (Figure 1.1b). The dipole moments are oriented differently among different ceramic grains, or even among different regions within a single grain. Regions of like-oriented dipole moments are referred to as domains, and each domain carries a net dipole moment. However, because the dipoles are randomly oriented in the ceramic element, as manufactured, the domains are randomly oriented and the element has no overall polarization (Figure 1.2a).

The ceramic element is polarized (poled) by exposing it to a strong, direct current (DC) electric field, usually, but not exclusively, at a temperature slightly below the Curie point (Figure 1.2b). Through this polarizing (poling) treatment, domains most nearly aligned with the electric field expand at the expense of domains that are not aligned with the field, and the ceramic lengthens in the direction of the field. When the electric field is removed most of the dipoles are locked into this configuration of near alignment (Figure 1.2c). This gives the material a permanent polarization, called the remanent polarization, and a permanent deformation (elongation) that make it anisotropic—its properties differ, according to the direction in which they are measured. Analogous to corresponding characteristics of ferromagnetic materials, a poled ferroelectric material exhibits hysteresis, and dielectric constants for these materials are very high, and temperature dependent. Figure 1.3 shows a

Figure 1.4 Generator and motor actions of a piezoelectric element

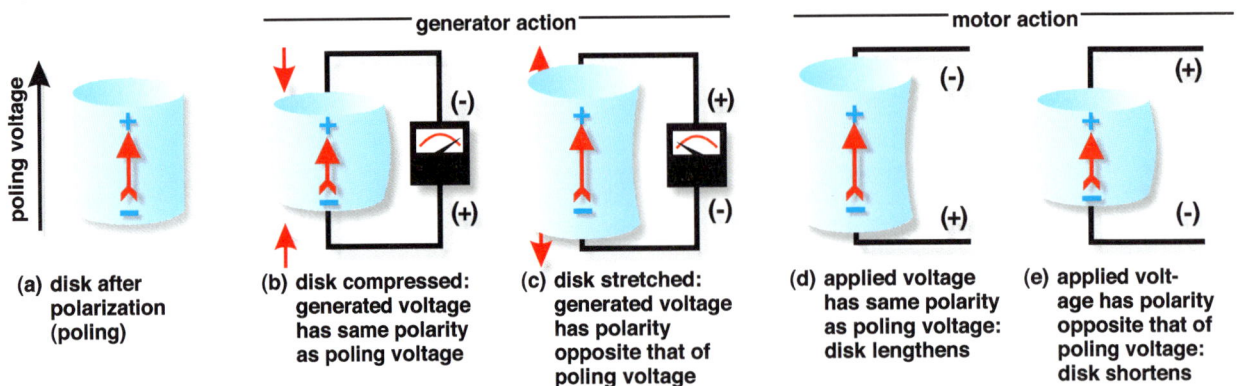

(a) disk after polarization (poling)

(b) disk compressed: generated voltage has same polarity as poling voltage

(c) disk stretched: generated voltage has polarity opposite that of poling voltage

(d) applied voltage has same polarity as poling voltage: disk lengthens

(e) applied voltage has polarity opposite that of poling voltage: disk shortens

Figure 1.5 Linear relationship between applied stress and generated voltage

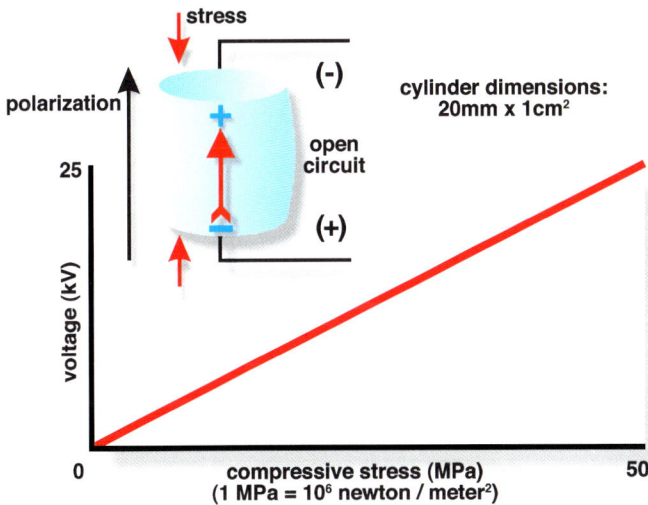

typical hysteresis curve created by applying an electric field to a piezoelectric ceramic element until maximum (saturation) polarization, P_s, is attained, reducing the field to zero to determine the remanent polarization, P_r, reversing the field to attain a negative maximum polarization and negative remanent polarization, and again reversing the field to restore the positive remanent polarization. The tracing below the hysteresis curve in Figure 1.3 plots the relative change in the dimension of the ceramic element along the direction of polarization, corresponding to the change in the electric field. The relative increase or decrease in the dimension parallel to the direction of the electric field is accompanied by a corresponding, but approximately 50% smaller, relative decrease or increase in the dimension perpendicular to the electric field.

Mechanical compression or tension on a poled piezoelectric ceramic element changes the dipole moment, creating a voltage. Compression of the element along the direction of polarization, or tension perpendicular to the direction of polarization, generates voltage of the same polarity as the poling voltage (Figure 1.4b). If tension is applied along the direction of polarization, or the element is compressed perpendicular to the

direction of polarization, the polarity of the voltage is opposite that of the poling voltage (Figure 1.4c). These actions are generator actions— the ceramic element converts the mechanical energy of compression or tension into electrical energy. This behavior has been adapted for use in fuel-igniting devices, solid state batteries, and other products.

If a voltage of the same polarity as the poling voltage is applied to a piezoelectric element, parallel to the direction of the poling voltage, the element will lengthen and its diameter will become smaller (Figure 1.4d). If a voltage of polarity opposite that of the poling voltage is applied, the element will become shorter and broader (Figure 1.4e). If an alternating voltage is applied, the disk will lengthen and shorten cyclically, at the frequency of the applied voltage. This is motor action—electrical energy is converted into mechanical energy. The principle is adapted to piezoelectric motors, sound or ultrasound generating devices, and many other products. Figure 1.5 shows that values for compressive stress and the voltage (or field strength) generated by applying stress to a piezoelectric ceramic element are linearly proportional up to a material-specific stress. The same is true for applied voltage and generated strain.

Equation 1.1

$$P = D - \varepsilon^T E$$

where

P = polarization
D = electric displacement
ε^T = permittivity of ceramic at constant stress
E = electric field

P also can be determined from the relationship $P = -(dT)$, in which d is the piezoelectric charge constant and T is the compressive stress on the ceramic element (eliminate the -sign for a tensile stress). Permittivity (dielectric constant), piezoelectric charge constants, and other piezoelectric constants are described in the next section of this chapter.

Table 1.1 Frequently-used constants

Piezoelectric Charge Constant d_{xy}

d_{33} — induced polarization in direction 3 (parallel to direction in which ceramic element is polarized) per unit stress applied in direction 3

or

induced strain in direction 3 per unit electric field applied in direction 3

d_{31} — induced polarization in direction 3 (parallel to direction in which ceramic element is polarized) per unit stress applied in direction 1 (perpendicular to direction in which ceramic element is polarized)

or

induced strain in direction 1 per unit electric field applied in direction 3

d_{15} — induced polarization in direction 1 (perpendicular to direction in which ceramic element is polarized) per unit shear stress applied about direction 2 (direction 2 perpendicular to direction in which ceramic element is polarized)

or

induced shear strain about direction 2 per unit electric field applied in direction 1

Piezoelectric Voltage Constant g_{xy}

g_{33} — induced electric field in direction 3 (parallel to direction in which ceramic element is polarized) per unit stress applied in direction 3

or

induced strain in direction 3 per unit electric displacement applied in direction 3

g_{31} — induced electric field in direction 3 (parallel to direction in which ceramic element is polarized) per unit stress applied in direction 1 (perpendicular to direction in which ceramic element is polarized)

or

induced strain in direction 1 per unit electric displacement applied in direction 3

g_{15} — induced electric field in direction 1 (perpendicular to direction in which ceramic element is polarized) per unit shear stress applied about direction 2 (direction 2 perpendicular to direction in which ceramic element is polarized)

or

induced shear strain about direction 2 per unit electric displacement applied in direction 1

Permittivity ε_{xy}

ε^T_{11} — permittivity for dielectric displacement and electric field in direction 1 (perpendicular to direction in which ceramic element is polarized), under constant stress

ε^S_{33} — permittivity for dielectric displacement and electric field in direction 3 (parallel to direction in which ceramic element is polarized), under constant strain

Elastic Compliance s_{xy}

s^E_{11} — compliance for stress in direction 1 (perpendicular to direction in which ceramic element is polarized) and accompanying strain in direction 1, under constant electric field (short circuit)

s^D_{33} — compliance for stress in direction 3 (parallel to direction in which ceramic element is polarized) and accompanying strain in direction 3, under constant electric displacement (open circuit)

k$_{33}$ factor for electric field in direction 3 (parallel to direction in which ceramic element is polarized) and longitudinal vibrations in direction 3 (ceramic rod, length >10x diameter)

k$_t$ factor for electric field in direction 3 and vibrations in direction 3 (thin disk, surface dimensions large relative to thickness; k$_t$ < k$_{33}$)

k$_{31}$ factor for electric field in direction 3 (parallel to direction in which ceramic element is polarized) and longitudinal vibrations in direction 1 (perpendicular to direction in which ceramic element is polarized) (ceramic rod)

k$_p$ factor for electric field in direction 3 (parallel to direction in which ceramic element is polarized) and radial vibrations in direction 1 and direction 2 (both perpendicular to direction in which ceramic element is polarized) (thin disk)

Piezoelectric Constants

Because a piezoelectric ceramic is anisotropic, physical constants relate to both the direction of the applied mechanical or electric force and to the directions perpendicular to the applied force. Consequently, each constant generally has two subscripts that refer to the directions of the two related quantities, such as stress (force on the ceramic element / surface area of the element) and strain (change in length of element / original length of element) for elasticity. The direction of positive polarization usually is made to coincide with the Z-axis of a rectangular system of X, Y, and Z axes (Figure 1.6). Direction X, Y, or Z is represented by the subscript 1, 2, or 3, respectively, and shear about one of these axes is represented by the subscript 4, 5, or 6, respectively. A superscript indicates a quantity kept constant. For example: ε^T_{11} is the permittivity for the dielectric displacement in direction 1 and electric field in direction 1 under constant stress, and ε^S_{33} is the permittivity for the dielectric displacement in direction 3 and electric field in direction 3 under constant strain. Definitions of important constants follow. Some of the most frequently used constants are summarized in Table 1.1. Equations for determining and interrelating these constants are summarized in Table 1.2.

Piezoelectric Charge Constant

The piezoelectric charge constant, d, is, alternatively, the polarization generated per unit of mechanical stress (T) applied to a piezoelectric

material or the mechanical strain (S) experienced by a piezoelectric material per unit of electric field applied (Table 1.2).

The first subscript to d indicates the direction of polarization generated in the material when the electric field, E, is zero or, alternatively, is the direction of the applied field strength. The second subscript is the direction of the applied stress or the induced strain, respectively. Because the extent of the strain induced in a piezoelectric material by an applied electric field is the product of the value for the electric field and the value for d (S = dE), d is an important value (i.e., a figure of merit) for ascertaining a material's suitability for strain-dependent (actuator) applications.

Figure 1.6 Directions of forces affecting a piezoelectric element

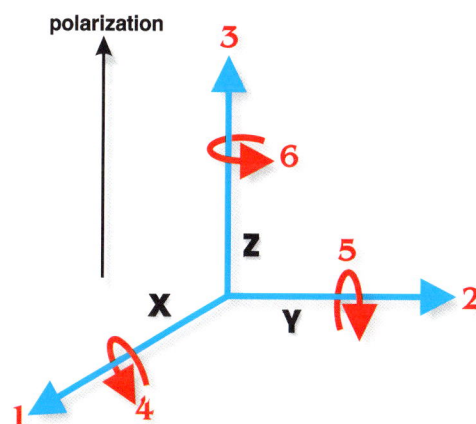

Table 1.2 Most-used constants and mathematical equations

Aging Rate

rate of aging $= (Par_2 - Par_1) / (Par_1 (\log t_2 - \log t_1))$

Bandwidth

$B \cong kf_p \text{ or } B \cong kf_s$

Dielectric Constant (Relative)

permittivity of ceramic material / permittivity of free space (8.85×10^{-12} farad / m)

$K^T = \varepsilon^T / \varepsilon_0$

Dielectric Dissipation Factor (Dielectric Loss Factor)

conductance / susceptance for parallel circuit equivalent to ceramic element; tangent of loss angle ($\tan \delta$) (measure directly, typically at 1 kHz)

Elastic Compliance

strain developed / stress applied; inverse of Young's modulus (elasticity)

$s \quad = 1 / \rho \upsilon^2$

$s^D_{33} = 1 / Y^D_{33}$

$s^E_{33} = 1 / Y^E_{33}$

$s^D_{11} = 1 / Y^D_{11}$

$s^E_{11} = 1 / Y^E_{11}$

Electromechanical Coupling Factor

mechanical energy converted / electrical energy input, or
electrical energy converted / mechanical energy input

Static / low frequencies

ceramic plate

$k_{31}^2 = d_{31}^2 / (s^E_{11} \varepsilon^T_{33})$

ceramic disk

$k_p^2 = 2d_{31}^2 / (s^E_{11} + s^E_{12}) \varepsilon^T_{33})$

ceramic rod

$k_{33}^2 = d_{33}^2 / (s^E_{33} \varepsilon^T_{33})$

Higher frequencies

ceramic plate

$$k_{31}^2 = \frac{(\pi / 2) (f_n / f_m) \tan [(\pi / 2) ((f_n - f_m) / f_m)]}{1 + (\pi / 2) (f_n / f_m) \tan [(\pi / 2) ((f_n - f_m) / f_m)]}$$

ceramic disk

$k_p \cong \sqrt{[(2.51 (f_n - f_m) / f_n) - ((f_n - f_m) / f_n)^2]}$

ceramic rod

$k_{33}^2 = (\pi / 2) (f_n / f_m) \tan [(\pi / 2) ((f_n - f_m) / f_n)]$

any shape

$k_{eff}^2 = (f_n^2 - f_m^2) / f_n^2$

Frequency Constant

(resonance frequency) (linear dimension governing resonance)

N_L = longitudinal mode $\quad N_L = f_s l$

N_p = radial mode $\quad N_p = f_s D_\phi$

N_T = thickness mode $\quad N_T = f_s h$

Mechanical Quality Factor

reactance / resistance for series circuit equivalent to ceramic element

$$Q_m = f_n^2 / (2\pi f_m C_0 Z_m (f_n^2 - f_m^2))$$

Piezoelectric Charge Constant

electric field generated by unit area of ceramic / stress applied, or
strain experienced by ceramic element / unit electrical field applied

$d = k\sqrt{(s^E \varepsilon^T)}$

$d_{31} = k_{31}\sqrt{(s^E_{11} \varepsilon^T_{33})}$

$d_{33} = k_{33}\sqrt{(s^E_{33} \varepsilon^T_{33})}$

$d_{15} = k_{15}\sqrt{(s^E_{55} \varepsilon^T_{11})}$

Piezoelectric Voltage Constant

electric field generated / stress applied, or
strain experienced by ceramic element / electric displacement applied

$g = d / \varepsilon^T$ $\quad g_{31} = d_{31} / \varepsilon^T_{33}$ $\quad g_{33} = d_{33} / \varepsilon^T_{33}$ $\quad g_{15} = d_{15} / \varepsilon^T_{11}$

Young's Modulus

stress applied / strain developed

$$Y = (F / A) / (\Delta l / l) = T/S$$

Symbols

A	surface area of ceramic element (m^2)	l	initial length of ceramic element (m)
B	bandwidth (frequency)	N	frequency constant (Hz·m)
d	piezoelectric charge constant (C / N)	Par_1	value for parameter Par at t_1 (days)
D_ϕ	diameter of ceramic disk or rod (m)	Par_2	value for parameter Par at t_2 (days)
ε_0	permittivity of free space (8.85×10^{-12} farad / m)	Q_m	mechanical quality factor
		ρ	density of ceramic (kg / m^3)
ε^T	permittivity of ceramic material (farad / m) (at constant stress)	s	elastic compliance (m^2 / N)
		S	strain
F	force	t_1	time 1 after polarization (days)
f_m	minimum impedance frequency (resonance frequency) (Hz)	t_2	time 2 after polarization (days)
		tan δ	dielectric dissipation factor
f_n	maximum impedance frequency (anti-resonance frequency) (Hz)	T	stress
		T°	temperature
f_p	parallel resonance frequency (Hz)	T_C	Curie point (°C)
f_s	series resonance frequency (Hz)	ν	velocity of sound in the ceramic material (m / s)
g	piezoelectric voltage constant (Vm / N)		
h	height (thickness) of ceramic element (m)	w	width of ceramic element (m)
k	electromechanical coupling factor	Y	Young's modulus (N / m^2)
k_{eff}	effective coupling factor	Z_m	minimum impedance at f_m (ohm)
K^T	relative dielectric constant (at constant stress)		

Permittivity

The permittivity, or dielectric constant, ε, for a piezoelectric ceramic material is the dielectric displacement per unit electric field. ε^T is the permittivity at constant stress, ε^S is the permittivity at constant strain. The first subscript to ε indicates the direction of the dielectric displacement; the second is the direction of the electric field.

The relative dielectric constant, K, is the ratio of ε, the amount of charge that an element constructed from the ceramic material can store, relative to the absolute dielectric constant, ε_0, the charge that can be stored by the same electrodes when separated by a vacuum, at equal voltage ($\varepsilon_0 = 8.85 \times 10^{-12}$ farad / meter).

Piezoelectric Voltage Constant

The piezoelectric voltage constant, g, is, alternatively, the electric field generated by a piezoelectric material per unit of mechanical stress applied or the mechanical strain experienced by a piezoelectric material per unit of electric displacement applied (Table 1.2). The first subscript to g indicates the direction of the electric field generated in the material, or the direction of the applied electric displacement. The second subscript is the direction of the applied stress or the induced strain, respectively. Because the strength of the induced electric field produced by a piezoelectric material in response to an applied physical stress is the product of the value for the applied stress and the value for g (Equation 1.2), g is a figure of merit for assessing a material's suitability for sensing (sensor) applications.

Equation 1.2

$$E = -(gT) + (D / \varepsilon^T)$$

where

E = electric field
T = stress applied parallel to axis of
    polarization* (T<O: compressive stress;
    T>O: tensile stress; see Figure 1.4)
D = electric displacement
ε^T = permittivity at constant stress
* applied force / surface area of element (N/m^2)

Equation 1.3 shows the relationship among d, ε^T, and g.

Equation 1.3

$$g = d / \varepsilon^T$$
or
$$d = g\varepsilon^T$$

Elastic Compliance

Elastic compliance, s, is the strain produced in a piezoelectric material per unit of stress applied and for the 11 and 33 directions, is the reciprocal of the modulus of elasticity (Young's modulus, Y) (Table 1.2). s^D is the compliance under a constant electric displacement; s^E is the compliance under a constant electric field. The first subscript indicates the direction of strain, the second is the direction of stress.

Young's Modulus

Young's modulus, Y, is an indicator of the stiffness (elasticity) of a ceramic material. Y is determined from the value for the stress applied to the material divided by the value for the resulting strain in the same direction (Table 1.2).

Electromechanical Coupling Factor

The electromechanical coupling factor, k, is an indicator of the effectiveness with which a piezoelectric material converts electrical energy into mechanical energy, or converts mechanical energy into electrical energy (Table 1.2). The first subscript to k denotes the direction along which the electrodes are applied; the second denotes the direction along which the mechanical energy is applied, or developed.

Under static or near-static conditions (input frequencies far below the resonance frequency of the piezoelectric material), either electrical to mechanical or mechanical to electrical conversion is expressed by:

Equation 1.4

$$k^2 = \text{converted (stored) energy / input energy}$$

At frequencies far below the resonance frequency, k can be determined for a ceramic plate, disk, or rod from equation 1.5, 1.6, or 1.7, respectively.

Equation 1.5 (for ceramic plate)

$$k_{31}^2 = d_{31}^2 / (s_{11}^E \varepsilon_{33}^T)$$

Equation 1.6 (for ceramic disk)

$$k_p^2 = 2d_{31}^2 / ((s_{11}^E + s_{12}^E)\varepsilon_{33}^T)$$

Equation 1.7 (for ceramic rod)

$$k_{33}^2 = d_{33}^2 / (s_{33}^E \varepsilon_{33}^T)$$

where

k = electromechanical coupling factor

k_p = electromechanical coupling factor for disk (surface dimensions large relative to thickness; see discussion below).

d = piezoelectric charge constant

s^E = elastic compliance (constant electric field)

ε^T = permittivity (constant stress)

At higher frequencies, k can be determined from equation 1.18, 1.19, or 1.20 (discussed later in this text, in **Dynamic Behavior**).

k values quoted in ceramic suppliers' specifications typically are theoretical maximum values. At low input frequencies, a typical piezoelectric ceramic can convert 30% to 75% of the energy stored in one form to the other form, depending on the formulation of the ceramic and the directions of the forces involved. A high k usually is desirable for efficient energy conversion, but k is not in itself a measure of efficiency, because it does not account for dielectric losses or mechanical losses. Further, unconverted energy often can be recovered. The true measure of efficiency is the ratio of converted, useable energy delivered by the piezoelectric element to the total energy taken up by the element. By this measure, piezoelectric ceramic elements in well designed systems can exhibit efficiencies that exceed 90%.

Figure 1.7 Radial (planar) vibrations in a piezoelectric ceramic disk

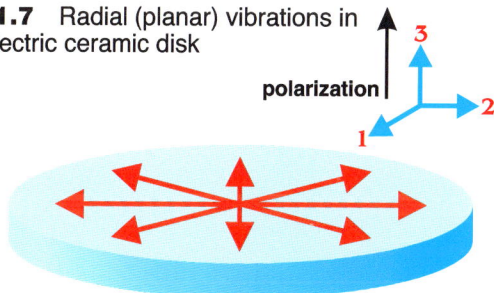

The dimensions of a ceramic element can dictate unique expressions of k: the planar coupling factor and the thickness coupling factor. For a thin disk of piezoelectric ceramic the planar coupling factor, k_p, expresses radial coupling—the coupling between an electric field parallel to the direction in which the ceramic element is polarized (direction 3) and mechanical effects that produce radial vibrations, relative to the direction of polarization (direction 1 and direction 2), as shown in Figure 1.7. For a disk or plate of material whose surface dimensions are large relative to its thickness, the thickness coupling factor, k_t, a unique expression of k_{33}, expresses the coupling between an electric field in direction 3 and mechanical vibrations in the same direction. The resonance frequency for the thickness dimension of an element of this shape is much higher than the resonance frequency of the transverse dimensions. At the same time, strongly attenuated transverse vibrations at this higher resonance frequency, a result of the transverse contraction / expansion that accompanies the expansion / contraction in thickness, give the element greater apparent stiffness. As a result, k_t is lower than k_{33}, the corresponding factor for longitudinal vibrations of a thin rod of the same material, for which a much lower longitudinal resonance frequency more closely matches the transverse resonance frequency.

Dielectric Dissipation Factor

The dielectric dissipation factor (dielectric loss factor), tan δ, for a ceramic material is the tangent of the dielectric loss angle. tan δ is determined by the ratio of effective conductance to effective susceptance in a parallel circuit, measured by using an impedance bridge. Values for tan δ typically are determined at 1 kHz.

Frequency Constant

When an unrestrained piezoelectric ceramic element is exposed to a high frequency alternating electric field, an impedance minimum, the planar or radial resonance frequency, coincides with the series resonance frequency, f_s. The relationship between the radial mode resonance frequency constant, N_p, and the diameter of the ceramic element, D_ϕ, is expressed by:

Equation 1.8
$$N_P = f_s D_\phi$$

At higher resonance, another impedance minimum, the axial resonance frequency, is encountered. The thickness mode frequency constant, N_T, is related to the thickness of the ceramic element, h, by:

Equation 1.9
$$N_T = f_s h$$

A third frequency constant, the longitudinal mode frequency constant, is related to the length of the element (Table 1.2).

It is helpful to remember that the velocity of sound in a piezoelectric ceramic element is approximately two times the frequency constant for the dimension controlling the output characteristics of the element.

Temperature Dependence of Piezoelectric Constants

The piezoelectric voltage constant, g, the piezoelectric charge constant, d, and the permittivity, ε, are temperature dependent factors. Within recommended operating temperature ranges, temperature-associated changes in the orientation of the domains usually are transient, but these changes can create charge displacements and electric fields that can obstruct measuring accuracy. Rarer, but possible, are sudden temperature fluctuations that generate relatively high voltages (see **Thermal Depolarization and Pyroelectric Effects**). These can depolarize the ceramic element or damage other components of the system.

If a capacitor is connected in parallel with the piezoelectric ceramic element, the increase in the total capacitance of the system will be accompanied by an equivalent reduction in the temperature coefficient for the total capacitance. Because $g = d / \varepsilon^T$ (Equation 1.3), the temperature coefficient for g is assumed to be the difference between the temperature coefficients for d and ε^T, and output voltage will be essentially constant over a wide temperature range.

Behavior of a Piezoelectric Ceramic Element

Low Frequency Input
At static or near-static input frequencies, the relationships between a force applied to a piezoelectric ceramic element and the electric field or charge produced are:

Equation 1.10
$$E = -(g_{33}T)$$
where

E = electric field
g_{33}= piezoelectric voltage constant*
T = stress on ceramic element
  * ceramic element subjected to mechanical stress parallel to direction of polarization, induced electric field in same direction

Equation 1.11
$$Q = -(d_{33}F)$$
where

Q = generated charge
d_{33}= piezoelectric charge constant*
F = applied force
  * ceramic element subjected to mechanical stress parallel to direction of polarization, induced electric field in same direction

The relationships between an applied voltage or electric field and the corresponding increase or decrease in a ceramic element's thickness, length, or width are:

Equation 1.12
$$\Delta h = d_{33}V$$

Equation 1.13
$$S = d_{33}E$$

Equation 1.14
$$\Delta l / l = d_{31}E$$

Dynamic Behavior (High Frequency Input)

A piezoelectric ceramic element exposed to an alternating electric field changes dimensions cyclically, according to the frequency of the field. Every ceramic element has a unique frequency, or resonance frequency, at which it vibrates most readily in response to the electrical input, and most efficiently converts the electrical energy input into mechanical energy. The resonance frequency is determined by the composition of the ceramic material and, in an inverse relationship, by the volume and shape of the element (generally, thicker elements have lower resonance frequencies).

Reactance of a ceramic element to a cyclic electrical input will vary as depicted in Figure 1.8. As the frequency of cycling increases, the oscillation frequency first approximates the series resonance frequency of the element—the frequency at which the impedance in an electrical circuit describing the element is zero when resistance caused by mechanical losses is ignored. At this frequency, the behavior of the element can be described by the equivalent electrical circuit in Figure 1.9. The frequency at which impedance is minimum, f_m, (maximum admittance; also called the resonance frequency, f_r) approximates the series resonance frequency, f_s:

As the frequency continues to increase, the impedance increases (Figure 1.8). The frequency at which the impedance becomes maximum, f_n, (minimum admittance; also called the anti-resonance frequency, f_a) approximates the parallel resonance frequency, f_p, the frequency at which parallel resistance in the equivalent electrical circuit becomes infinite when resistance caused by mechanical losses is ignored:

Figure 1.8 Impedance as a function of frequency

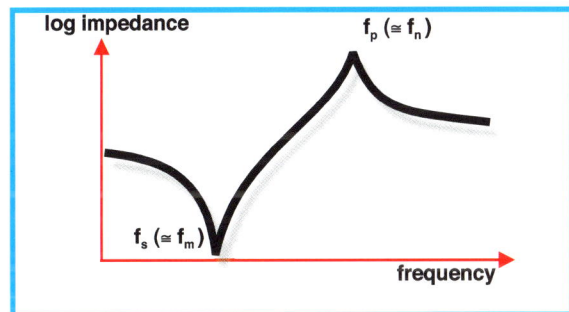

Figure 1.9 Electrical circuit equivalent to a piezoelectric ceramic transducer

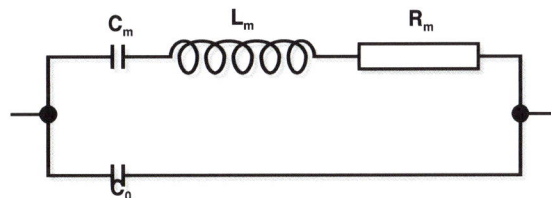

C_0 = (capacitance of transducer below resonance frequency) - (capacitance C_m)

C_m = capacitance of mechanical circuit

L_m = inductance of mechanical circuit

R_m = resistance caused by mechanical losses

Equation 1.17

$$f_p = (1 / 2\pi) (\sqrt{[(C_0 + C_m)/ L_m C_0 C_m]})$$
where
C_0 = (capacitance of ceramic element at
   resonance frequency) - (capacitance C_m)
C_m = capacitance of mechanical circuit
L_m = inductance of mechanical circuit

Below f_m and above f_n, the ceramic element behaves capacitively, between f_m and f_n, it behaves inductively, and the maximum response from the element will be at a point between f_m and f_n.

Values for f_m and f_n can be determined by experiment. Figure 1.10 shows a system designed to ascertain these values, and summarizes the measurement procedures.

The values for minimum impedance frequency, f_m, and maximum impedance frequency, f_n, can be used to calculate the electromechanical coupling factor, k. k depends on both the mode of vibration and the shape of the ceramic element. Equation 1.18, 1.19, or 1.20 expresses the relationship between k and f_m and f_n for a ceramic plate, a disk, or a rod, respectively.

Equation 1.18 (for ceramic plate)

$$k_{31}^2 = \frac{(\pi/2) (f_n/f_m) \tan [(\pi/2) ((f_n - f_m) / f_m)]}{1 + (\pi/2) (f_n/f_m) \tan [(\pi/2) ((f_n - f_m)/f_m)]}$$

Equation 1.19 (for ceramic disk - surface dimensions large relative to thickness)

$$k_p \approx \sqrt{[(2.51 (f_n - f_m) / f_n) - ((f_n - f_m) / f_n)^2]}$$

Equation 1.20 (for ceramic rod)

$$k_{33}^2 = (\pi/2) (f_n/f_m) \tan [(\pi/2) ((f_n - f_m)/f_n)]$$
where
k_{31} = transverse electromechanical
   coupling factor*
k_p = planar electromechanical
   coupling factor**
k_{33} = longitudinal electromechanical
   coupling factor***

* plate subjected to electric field parallel to
   direction of polarization, induced strain
   perpendicular to direction of polarization
** disk subjected to electric field parallel to
   direction of polarization, induced strain in
   same direction
*** rod subjected to electric field parallel to
   direction of polarization, induced strain in
   same direction

Figure 1.10 System for determining minimum impedance (resonance frequency) and maximum impedance (anti-resonance frequency) of a piezoelectric ceramic element

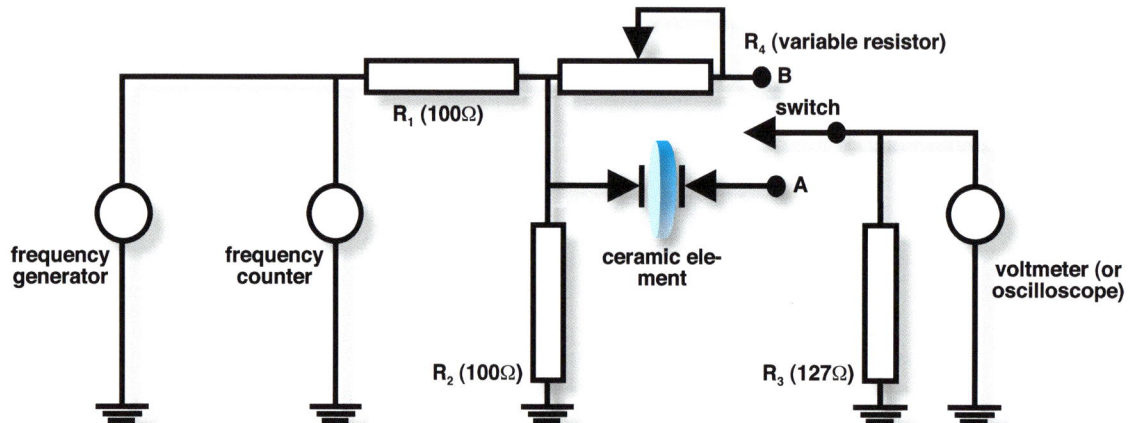

Procedure
1. Set switch to A.
2. Place ceramic element into position.
3. Adjust frequency generator to give maximum voltage value on voltmeter. This is f_m (resonance frequency).
4. Set switch to B.
5. Adjust R_4 to give voltage value on voltmeter equal to value in step 3. This R_4 value is the impedance resonance (Zr).
6. Set switch to A.
7. Adjust frequency generator to give minimum voltage value on voltmeter. This is f_n (anti-resonance frequency).

In addition to the electromechanical coupling factor, dielectric losses and mechanical losses affect efficiency of energy conversion. Dielectric losses, expressed by the dielectric dissipation factor, $\tan \delta$, usually are more significant than mechanical losses. $\tan \delta$ is measured directly, by using a capacitance bridge. From $\tan \delta$ the electric quality factor, Q_e, is defined by:

Equation 1.21

$$Q_e = 1 / \tan \delta$$

Mechanical loss is determined by calculating the mechanical quality factor, Q_m:

Equation 1.22

$$Q_m = f_n^2 / ((2\pi f_m C_0 Z_m)(f_n^2 - f_m^2))$$

where

f_n = frequency of impedance maximum
f_m = frequency of impedance minimum
C_0 = capacitance of ceramic element at low frequency
Z_m = minimum impedance

Mechanical loss, $\tan \delta_m$, is Q_m^{-1}.

A low Q_m indicates a ceramic material is a good harmonic oscillator. Materials with high Q_m are needed in applications in which maximum displacement and minimal heat generation are critical, such as for piezoelectric ultrasonic motors. Q_m is important for determining the magnitude of the induced strain at resonance frequency: the amplitude of the strain is proportional to the Q_m value (e.g., for a longitudinally vibrating rectangular ceramic plate maximum displacement is equal to $(\pi^2 / 8)(Q_m d_{31} El)$, where d is the piezoelectric charge constant, E is the value for the electric field, and l is the length of the plate).

At high frequencies Q_e and Q_m are not constants, but their values usually are smaller than at lower frequencies.

Tuning the piezoelectric element with a parallel inductance or a series inductance produces two coupled resonance circuits, one mechanical, the other electrical. The mechanical circuit is governed by the inductance of the mechanical circuit, L_m, the capacitance of the mechanical circuit, C_m, and the resistance caused by mechanical losses, R_m; the electrical circuit is governed by the parallel inductance, L_p, or the series inductance, L_s, the low frequency capacitance of the ceramic element, C_0, and the resistance of the alternating voltage generator, R. Parallel inductance L_p and series inductance L_s are expressed as:

Figure 1.11 Electrical circuits equivalent to tuned piezoelectric ceramic transducers

(a) parallel tuning

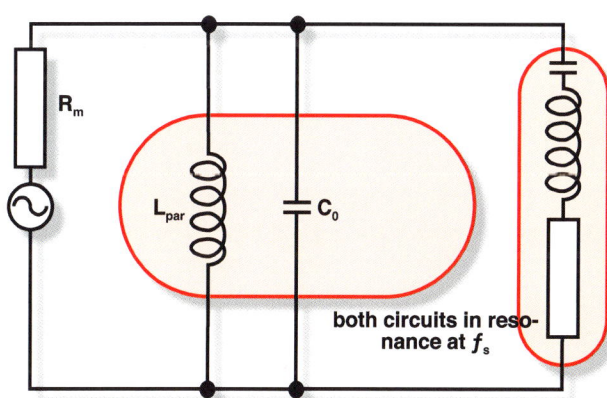

(b) series tuning

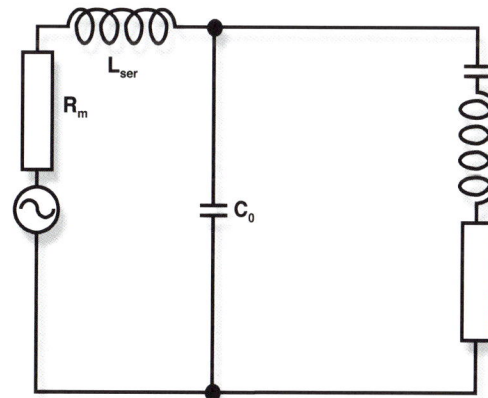

both circuits in resonance at f_s

C_0 = (capacitance of transducer below resonance frequency) - (capacitance of mechanical circuit)
R_m = resistance caused by mechanical losses
L_{par} = Inductance of mechanical circuit
L_{ser} = Inductance of mechanical circuit

Equation 1.23

$$L_p = 1 / ((2\pi f_p)^2 C_0)$$

Equation 1.24

$$L_s = 1 / ((2\pi f_s)^2 C_0)$$

where

f_p = parallel resonance frequency $(2\pi f_p = \omega_p)$
f_s = series resonance frequency $(2\pi f_s = \omega_s)$
C_0 = capacitance of ceramic element
    at low frequency

The electrical circuits in Figure 1.11 are equivalents for the tuned systems.

The signal bandwidth is approximately equal to the product obtained by multiplying the electromechanical coupling factor for the ceramic and the element parallel or series resonance frequency for the parallel-tuned or series-tuned system (Table 1.2).

Stability

Most properties of a piezoelectric ceramic element erode gradually, in a logarithmic relationship with time after polarization:

Equation 1.25

rate of aging = $(Par_2 - Par_1)/(Par_1 (\log t_2 - \log t_1))$

where

t_1 = time 1 after polarization (days)
t_2 = time 2 after polarization (days)
Par_1 = value for parameter Par at t_1
Par_2 = value for parameter Par at t_2

Exact rates of aging depend on the composition of the ceramic element and the manufacturing process used to prepare it.

Most ceramics age at two distinctly different rates. In the first 24-50 hours after polarization, losses of piezoelectric properties are significant, then this sharp decline is followed by a much slower degradation over the lifetime of the ceramic. From the example illustrated in Figure 1.12, the aging rate of this ceramic was calculated to be -2.4% per decade from immediately after polarization to approximately 1000-4000 minutes after polarization, and -0.9% per decade thereafter. The initial losses are attributed to some of the domains in the ceramic

Figure 1.12 Capacitance losses are greatest immediately after polarization

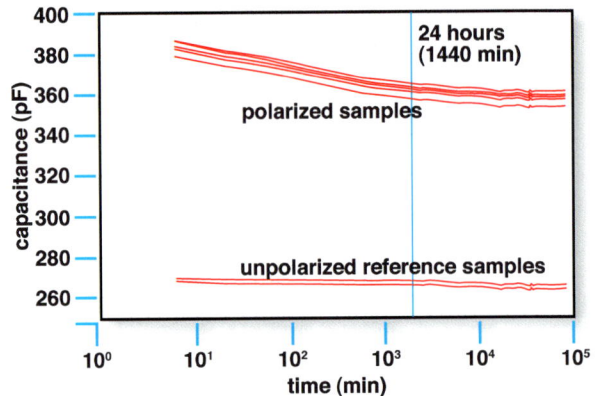

Material: Navy Type I ceramic ring,
15 mm OD x 6 mm ID x 5 mm high
Monitoring period: 60 days
Temperature constant for each sample (20°C - 26°C).

reverting to their original orientations relatively quickly after polarization. The subsequent slower degradation is governed by chemical defects within the grains and in the grain boundaries. This explanation is consistent with observations that high purity ingredients and careful powder production give a ceramic a more homogenous composition and fewer defects attributable to impurities. The aging rate for some Navy Type I materials, for example, have been determined to be up to 5 times lower than rates for other Navy Type I materials.

Mishandling the element can accelerate this inherent aging process: a piezoelectric ceramic element potentially can be depolarized by exceeding the electrical, mechanical, or thermal limitations of the element. Depolarization would reduce or destroy the piezoelectric properties.

Electrical Depolarization

Exposure to a strong electric field of polarity opposite that of the polarizing field will depolarize a piezoelectric material. The degree of depolarization depends on the grade of material, the exposure time, the temperature, and other factors, but fields of 200-500 V/mm or greater typically have a significant depolarizing effect. An alternating current will have a depolarizing effect during each half cycle in which polarity is opposite that of the polarizing field.

Mechanical Depolarization

Mechanical stress sufficient to disturb the orientation of the domains in a piezoelectric material can destroy the alignment of the dipoles. Like susceptibility to electrical depolarization, the ability to withstand mechanical stress differs among the various grades and brands of piezoelectric materials.

Thermal Depolarization and Pyroelectric Effects

If a piezoelectric ceramic material is heated to its Curie point, the domains will become disordered and the material will be depolarized. The recommended upper operating temperature for a ceramic usually is approximately half-way to slightly more than half-way between 0°C and the Curie point.

Within the recommended operating temperature range, temperature-associated changes in the orientation of the domains are reversible, but these changes can create charge displacements and electric fields. Also, sudden temperature fluctuations can generate relatively high voltages, capable of depolarizing the ceramic element. In most applications these are unwanted effects that can obstruct measuring accuracy and damage or destroy sensitive electronic components. To suppress the effects of heat or temperature fluctuation, a parallel capacitor can be incorporated into the system, to accept the additional electrical energy.

For a particular ceramic material, two values, the pyroelectric charge constant, $\Delta P_{pyr}/\Delta T^\circ$, which quantifies the change in polarization for a given change in temperature, and the pyroelectric field strength constant, $\Delta E/\Delta T^\circ$, which quantifies the change in electric field for a given change in temperature, are indicators of the vulnerability of the material to pyroelectric effects. Resistance to pyroelectric effects is strongest for a ceramic with a high ratio of piezoelectric charge constant (d_{33}) to pyroelectric charge constant or, alternatively, a high ratio of piezoelectric voltage constant (g_{33}) to pyroelectric field strength constant. The pyroelectric field strength constant is useful for predicting the behavior of a ceramic under open circuit conditions (e.g., in a fuel ignitor).

For a given temperature change the pyroelectric charge, Q_p, can be calculated from:

Equation 1.26

$$Q_p = 1/2\ ((\Delta P_{pyr}/\Delta T^\circ_1 + \Delta P_{pyr}/\Delta T^\circ_2)\ (T^\circ_2 - T^\circ_1)\ A)$$
where
$\Delta P_{pyr}/\Delta T^\circ_1$ = pyroelectric charge constant at temperature 1
$\Delta P_{pyr}/\Delta T^\circ_2$ = pyroelectric charge constant at temperature 2
A = surface area of ceramic element

The pyroelectric voltage generated by a ceramic element, V_p, is:

Equation 1.27

$$V_p = (\Delta E/\Delta T^\circ)\ (\Delta T^\circ)\ (h)$$
where
$\Delta E/\Delta T^\circ$ = change in electric field
ΔT° = change in temperature
h = height (thickness) of ceramic element

Typical Applications

A piezoelectric system can be constructed for virtually any application for which any other type of electromechanical transducer is used. For any particular application, however, the limiting factors are the size, weight, and cost of the system. Some of the most widely used applications for piezoceramic materials manufactured by APC International, Ltd. are briefly summarized here. Subsequent chapters of this publication contain more detailed information about these applications.

Generators

Piezoelectric ceramics can generate voltages sufficient to create a spark across an electrode gap, and thus can be used as ignitors in fuel lighters, gas stoves, welding equipment, and other such appliances. A piezoelectric ignition system is smaller and far less complex than alternative systems using permanent magnets or high voltage transformers and capacitors.

Techniques used to make multilayer capacitors can be adapted to make multilayer piezoelectric ceramic generators. A large surface area per unit volume makes for a high generated charge and relatively low voltage. Such generators are excellent solid state batteries for electronic circuits.

Sensors

Sensors convert various physical parameters, such as acceleration or pressure, into electrical signals. In some types of sensors the physical parameter being monitored acts directly on the piezoelectric ceramic element; in other devices an acoustical signal establishes vibrations in the ceramic element and these vibrations are, in turn, converted into an electrical signal.

Actuators

A piezoelectric actuator converts a voltage or other electrical signal into precisely controlled physical displacement that can be used to finely adjust machining tools, or lenses or mirrors in optical equipment. Alternatively, this displacement can be prevented, enabling a useable force to develop. In addition to position control applications, piezoelectric actuators are used to actuate or control hydraulic valves, to act as small-volume pumps or special-purpose motors, and in other applications. Energy efficiency losses limit the extent to which electromagnetic motors can be reduced in size. Piezoelectric motors, on the other hand, are unaffected by these constraints—energy efficiency is independent of size. This makes it possible to design piezoelectric motors to sizes of less than 1 cm^3. An important additional advantage, in some applications, is the fact that piezoelectric motors do not generate electromagnetic noise.

Transducers

Piezoelectric transducers convert electrical energy into vibrational mechanical energy, usually expressed as sound or ultrasound. The mechanical energy, in turn, is used to perform a task.

Transducers that generate audible sounds afford significant advantages, relative to comparable electromagnetic devices. They are compact, simple, and highly reliable, and they require

Table 1.3

Characteristics of piezoelectric ceramics compared

Characteristic	Soft Ceramic	Hard Ceramic
Piezoelectric Constants	larger	smaller
Permittivity	higher	lower
Dielectric Constants	larger	smaller
Dielectric Losses	higher	lower
Electromechanical Coupling Factors	larger	smaller
Electrical Resistance	very high	lower
Mechanical Quality Factors	low	high
Coercive Field	low	higher
Linearity	poor	better
Polarization / Depolarization	easier	more difficult

minimal energy to produce a high level of sound. These characteristics are ideally matched to the needs of battery-powered equipment.

Transducers that generate ultrasonic vibrations are used for cleaning, atomizing liquids, drilling or milling ceramics or other difficult materials, welding plastics, medical diagnostics, and for many other purposes. Because the piezoelectric effect is reversible, a transducer can both generate an ultrasound signal and receive a reflected signal. Consequently, systems incorporating either a single piezoelectric transducer or two transducers are widely used to measure distances and flow rates, and as fluid level sensors (by differentiating between sound transmission in the fluid and in air).

Soft Ceramics versus Hard Ceramics

A piezoelectric ceramic is categorized as a "soft" material or a "hard" material according to the nature bestowed on it by its composition. Soft ceramics and hard ceramics contain "donor" dopants or "acceptor" dopants, respectively. Small amounts of a donor dopant added to a lead zirconate titanate formulation create metal (cation) vacancies in the crystal structure (1), thereby enhancing the reorientation of domains when the ceramic is exposed to a polarizing electric field. The soft PZTs produced from these formulations are characterized by large electromechanical coupling factors, large piezoelectric constants, high permittivity, large dielectric constants, high dielectric losses, low mechanical quality factors, and poor linearity (Table 1.3). Soft ceramics produce large displacements, relative to hard ceramics, but they exhibit greater hysteresis than hard ceramics, and are more susceptible to depolarization or other deterioration. Lower Curie points (generally below 300°C) dictate that soft ceramics be used at lower temperatures. Generally, large values for permittivity and dielectric dissipation factor (tan δ) restrict or eliminate soft ceramics from consideration for applications requiring combinations of high frequency inputs and high electric fields. Consequently, soft ceramics are used primarily in sensing applications, rather than in power applications.

Acceptor dopants in a ceramic formulation create oxygen (anion) vacancies in the crystal structure (1). Hard PZTs have characteristics generally opposite those of soft ceramics (Table 1.3). They also are more difficult to polarize or depolarize. Hard ceramics generally are more stable than soft ceramics, but they cannot produce the same large displacements. Hard ceramics have Curie points above 300°C. They exhibit small values for the piezoelectric charge constants (d) and large values for mechanical quality factor (Q_m). Hard ceramics are compatible with high mechanical loads and high voltages.

Ceramic elements

Note that these statements are generalizations. A soft ceramic can be prepared to exhibit some characteristics approaching those of a hard ceramic, or *vice versa*. Thus, when choosing a ceramic for a particular application, it is helpful to look beyond general categorization, and carefully compare specific characteristics.

New Approaches

Piezoelectric ceramic materials are being used in a wide and increasing range of applications. In any application, however, there is always room for improved performance: greater movement, longer useful lifetime, higher temperature limits, etc. In recent decades many investigators have begun to evaluate new approaches to employing the piezoelectric effect. The following are among the currently active subjects of research.

Materials with Diffuse Phase Transition (Relaxors)

In conventional piezoelectric materials, piezoelectric behavior disappears above a sharply defined critical temperature, the Curie point, because at and above this point each crystal exhibits a simple cubic symmetry with no dipole moment. At temperatures below the Curie point, each crystal has a tetragonal or rhombohedral symmetry and a dipole moment, and thus can display piezoelectric characteristics. In relaxor materials, in contrast, the transition between capability for piezoelectric behavior and loss of piezoelectric capability does not occur at a specific temperature, but instead occurs over a temperature range, the Curie range. Within the Curie range is a temperature at which electrical permittivity is maximum. Further complicating this situation is the fact that the electrical permittivity maximum also is frequency dependent, and is displaced to higher temperatures as the frequency of an applied input is increased (within the radiofrequency spectrum). This behavior produces a disperse curve relating permittivity and temperature. Relative insensitivity to temperature and the very high electromechanical coupling factors exhibited by single crystals of some relaxor formulations (>0.9, versus factors of 0.7 - 0.8 for conventional PZT ceramics) make relaxors very attractive materials for actuator, transducer, and other applications. Lead magnesium niobate, lead magnesium niobate doped with varying amounts of lanthanum, and lead nickel niobate currently are among the most studied relaxors.

Table 1.4 Physical and performance properties of widely used single-crystal materials

β-BaB$_2$O$_4$ (BBO)

Density (g/cm^3):	3.85
Refractive Index:	n$_0$: 1.6749; n$_e$: 1.5555
Transmission Range (μm):	0.190 - 3.5
Thermal Expansion Coefficient (K^{-1}):	a: 4.0 x 10^{-6}; c: 2.0 x 10^{-6}

Bi$_4$Ge$_3$O$_{12}$ (BGO)

Crystal Cell (Å):	10.518
Density (g/cm^3):	7.13
Melting Point (°C):	1050
Refractive Index:	2.15
Radiation Length (cm):	1.1
Peak of Fluorescence Spectrum (nm):	480
Delay Time (ns):	300
Relative Light Output (%):	10 - 14 NaI (T1)
Energy Resolution (%, at 511keV):	20

Li$_2$B$_4$O$_7$ (LBO)

Point Group:	4mm
Lattice Constant (Å):	a: 9.475; c: 10.283
Density (g/cm^3):	2.45
Hardness (Mohs):	~6
Melting Point (°C):	917
Solubility (M/h):	pure water: a: 0.04; c: 0.08
	conc. HNO$_3$: 60
	KOH: 0.4
	CHCl$_3$: 0
Dielectric Constant:	$\varepsilon_{11}/\varepsilon_0$: 3.5; $\varepsilon_{33}/\varepsilon_0$: 8.2
Piezoelectric Charge Constant (C/N):	d$_{15}$: 8.07 x 10^{-12}; d$_{33}$: 19.4 x 10^{-12}
Temperature Coefficient of Delay(K^{-1}, at 25°C):	0 x 10^{-6}
Thermal Expansion Coefficient (K^{-1}):	a$_{11}$: 11.1 x 10^{-6}; a$_{33}$: 3.7 x 10^{-6}
SAW Coupling Factor (k^2):	0.7 - 1.6
SAW Velocity (m/s):	3402 - 3864

LiNbO$_3$

Density (g/cm^3):	4.64
Hardness (Mohs):	5
Melting Point (°C):	1250
Curie Temperature (°C):	1140

LiTaO$_3$

Crystal Structure:	trigonal, space group R$_{3c}$, point group 3m
Cell Parameters (Å):	a: 5.154.; c: 13.781
Density (g/cm^3):	7.46
Hardness (Mohs):	5.5
Melting Point (°C):	1650
Refractive Index (at 632.8nm):	n$_0$: 2.176; n$_e$: 2.180
Electro-Optical Coefficient (pm/V):	R$_{33}$: 30.4
Transmission Range (nm):	400 - 4500
Dielectric Constant:	$\varepsilon_{11}/\varepsilon_0$: 51.7; $\varepsilon_{33}/\varepsilon_0$: 44.5

LiTaO₃ *(cont'd)*

Piezoelectric Strain Constant (C/N): d_{22}: 2.4×10^{-11}; d_{33}: 0.8×10^{-11}
Elastic Stiffness Coefficient (N/m²): c^E_{11}: 2.33×10^{11}; c^E_{33}: 2.77×10^{11}
Curie Temperature (°C): 607

PMN-PT

Density (g/cm³): 7.64
Electromechanical Coupling Coefficients: k_{31}: 0.55; k_{33}: 0.94; K_t: 0.95
Relative Dielectric Constant: 5000
Frequency Constants (Hz•m): N_{31}: 2100; N_{33}: 3200; N_T: 2800
Elastic Constants (N/m²): Y^E_{11}: 8.7×10^{10}; Y^E_{33}: 8.2×10^{10}
Piezoelectric Charge Constant (pC/N): d_{31}: -600; d_{33}: 1500
Phase Change Temperature (°C): 70
Dissipation Factor (%): 0.56
Curie Temperature (°C): 150

PbMoO₄

Color: light yellow
Density (g/cm³): 6.95
Transmission (%, at 0.63μm): 488nm: 67.7; 632.8nm: 71.6
Acoustic Velocity (mm/μs): <110>, <001>: 3.63
Acoustic/Optic Figure of Merit (s³/g): <110>, <001>: 36.3×10^{-18}

TeO₂

Density (g/cm³): 6
Transmission (%, at 0.63μm): 70
Acoustic Velocity (mm/μs): <110>: 0.62; <001>: 4.2
Acoustic/Optic Figure of Merit (s³/g): <110>: 1200×10^{-18}; <001>: 34.5×10^{-18}

YVO₄

Crystal Class: positive uniaxial, $n_0 = n_a = n_b$, $n_e = n_c$
Crystal Symmetry: zircon tetragonal, space group D4h
Crystal Cell (Å): a,b: 7.12; c: 6.29
Walk-Off Angle at 45° (ρ) 0.63mm: 6.04°; 1.30mm: 5.72°; 1.55mm: 5.69°
Seamier Equation (λ in μm): n_0^2: $3.77834 +$
$0.069736/(\lambda^2 - 0.04724) - 0.0108133\lambda^2$
n_e^2: $4.59905 +$
$0.110534/(\lambda^2 - 0.04813) - 0.0122676\lambda^2$
Density (g/cm³): 4.22
Hardness (Mohs): 5 (glass-like)
Transparency Range: high transmission from 0.4 μm to 5 μm
Hygroscope Susceptibility: non-hygroscopic
Refractive Indices, Birefringence: 0.63mm: n_0: 1.9929, n_e: 2.2154, Δ_n: 0.2225
$(\Delta_n = n_e - n_0)$ 1.30mm: n_0: 1.9500, n_e: 2.1554, Δ_n: 0.2054
1.55mm: n_0: 1.9447, n_e: 2.1486, Δ_n: 0.2039
Thermal Expansion Coefficient (K⁻¹): αa: 4.43×10^{-6}; αc: 11.37×10^{-6}
Thermal Conductivity Coefficient (W/m/K): //C: 5.23; ⊥C:5.10
$\Delta n_a /\Delta T$: 8.5×10^{-6}/K; $\Delta n_c /\Delta T$: 3.0×10^{-6}/K

Single-Crystal Piezoelectric Elements

A piezoelectric ceramic element is a mass of regions of local dipole alignment (domains) locked in near-alignment with the direction of the electric field in which the element was polarized. It is reasonable to conclude that the piezoelectric properties of a ceramic element would be improved if all of the domains in the element were perfectly aligned. In the current state of ceramics technology such a ceramic cannot be produced, but single natural crystals of materials with piezoelectric behavior can be obtained, and a wide and expanding variety of man-made single crystals is being developed. These crystals are used in acoustical, optical, wireless communication, and other applications.

Materials currently used to fabricate single-crystal piezoelectric elements include lead magnesium niobate / lead titanate (PMN-PT), lead zirconate niobate / lead titanate (PZN-PT), lithium niobate ($LiNbO_3$), lithium niobate with dopants, lithium tetraborate ($Li_2B_4O_7$), and quartz. Single-crystal PMN-PT and PZN-PT elements exhibit ten times the strain of conventional polycrystalline PZT elements. Applications for these single-crystal materials include actuators and both diagnostic and invasive medical devices. A useful combination of piezoelectric and electro-optic properties makes lithium niobate and doped lithium niobate crystals very useful for surface acoustic wave (SAW) devices and electro-optical applications. Lithium tetraborate crystals exhibit a fairly high electromechanical coupling factor (k); a SAW chip made from this material can be 60% smaller than the lithium niobate or quartz alternative. In addition to SAW devices, applications for lithium tetraborate crystals include bulk acoustic wave (BAW) devices, pagers, cordless and cellular telephones, and data communication devices. Applications for quartz crystals include timing mechanisms for watches and clocks and delay lines for electrical circuits.

Barium titanate ($BaTiO_3$) is a potential non-lead source of piezoelectric crystals for low temperature and room temperature applications. At 0°C, electromechanical coupling factor (k_{33}) and piezoelectric charge constant (d_{33}) values for $BaTiO_3$ single crystals were approximately 85% and approximately 500 pC/N, respectively (2). At -90°C, corresponding values were approximately 79% and approximately 400 pC/N, respectively. In comparison, room temperature k_{33} values for polycrystalline lead zirconate titanate materials range between 50% and 75%, and upper-level values for d_{33} are 600-700 pC/N.

A number of other materials also are being used to develop single crystals for optical and/or piezoelectric applications. beta-barium borate (β-BaB_2O_4 / BBO) crystals exhibit a large birefringence over a broad transparent range, from 189 nm to 3500 nm, and have stable physical and chemical properties and a high damage threshold. BBO is an excellent replacement for calcite ($CaCO_3$), rutile (TiO_2), lithium niobate ($LiNbO_3$), or other crystals in Glan prisms, and in high-power or UV wavelength applications.

Bismuth germanate ($Bi_4Ge_3O_{12}$ / BGO) is the crystalline form of a transparent, colorless, water-insoluble inorganic oxide with cubic eulytine structure. When exposed to radiation of high-energy particles or other sources, such as gamma rays or x-rays, it emits a green fluorescent light with a peak wavelength of 480 nm. High stopping power, high scintillation efficiency, good energy resolution, and a non-hygroscopic nature make BGO a good scintillation material that has found a wide range of applications, including high energy physics, nuclear physics, space physics, nuclear medicine, and geological prospecting.

Lithium tetraborate crystals ($Li_2B_4O_7$ / LBO) exhibit excellent piezoelectric properties, and applications for this material have been widely and extensively developed. It exhibits low density, a weak temperature coefficient of delay (TCD), and a high SAW coupling factor (k^2). It is not necessary to polarize the crystals before use. LBO is ideal for use in SAW filters, resonators, and other high-frequency devices.

Lithium niobate with dopants has useful properties. $LiNbO_3$ with 0.01 to 0.1 mol % Fe is used for erasable volume phase holographic storage. $LiNbO_3$ with 1 to 6 mol % MgO is used for frequency doubling and optic parametric

Figure 1.13 Polyvinylidene difluoride (PVDF)

(CH₂CF₂)ₙ

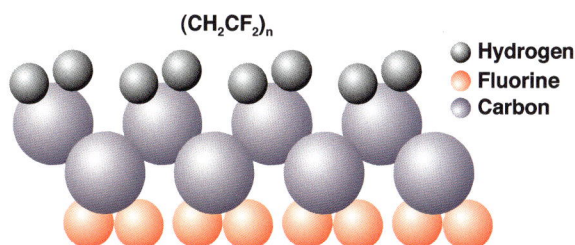

- ● Hydrogen
- ● Fluorine
- ● Carbon

oscillation applications. $LiNbO_3$ with Mg / Nd is used for self-frequency doubled, Q-switched, or modulated lasers.

Lithium tantalate crystals ($LiTaO_3$) have optic, NLO, and electro-optic properties similar to those of $LiNbO_3$ crystals, but with a higher damage threshold.

Good acoustic-optic materials, lead molybdenum oxide crystals ($PbMoO_4$) have been used in A-D modulator, deflector, and tunable filter applications.

Tellurium oxide (TeO_2) crystals have numerous applications. When an electrical signal is applied to a piezoelectric transducer bonded to an anisotropic medium to generate ultrasonic waves (propagated through the medium) a three-dimensional grating is formed by virtue of the refractive index variation associated with the travelling acoustic wave. The incident beam passing through the medium will be coupled into several symmetrical diffraction orders, as in a two-dimensional diffraction grating. TeO_2 crys-

tals are used in acoustic-optic modulators and deflectors, and acoustic-optic tunable filters with wavelengths ranging from 0.8 μm to 5.0 μm are being fabricated for use in laser printers, signal processing, high speed optical beam deflectors, RF spectrum analysis, rapid-scan spectrometers and radiometers, and other applications.

Yttrium orthovanadate (YVO_4) is ideal for optical polarizing components because of its wide transparency range and large birefringence. It is an excellent synthetic substitute for calcite or rutile crystals in many applications, including fiber optic isolators, circulator beam displacers, and other polarizing optics.

Typical physical and performance characteristics of several single-crystal materials are summarized in Table 1.4.

The performance of a single crystal piezoelectric element depends on the alignment of the crystal relative to the applied electric field (i.e., the direction in which the raw crystal is cut). The crystal can be cut in such a way that the deformation created by applying an electric field will maximize either thickness expansion or shear: a cut normal to the x axis produces an element with maximal potential for expanding in thickness; a cut normal to the y axis will maximize the potential for shear distortion.

Similarly, a polycrystalline ceramic cut at an angle of 57° to the direction of polarization will exhibit superior piezoelectric properties.

Table 1.5 Piezoelectric properties of polyvinylidene difluoride (PVDF) and piezoelectric ceramics*

Property	Units	PVDF	BaTiO₃	PZT
Dielectric Constant		12	600-1200	<1000-4000
Piezoelectric Charge Constant	pC/N	$d_{31} = 20$	-30 - -60	~-100 - >-600
		$d_{33} = -30$	<100-150	~ 200 - > 600
Electromechanical Coupling Factor	%	11	21	30-75
Young's Modulus	10^{10} N/m²	0.3	11-12	6-9
Acoustic Impedance	10^6 kg/(m²s)**	2.3	25	30
Density	kg/m³	1780	5300-5700	7500-7700

* At 20°C.

** MRa (megarayleigh) (1 Ra = 1 kg/(m²s)).

PVDF data from Fraden, J., Handbook of Modern Sensors (2nd ed.), American Institute of Physics, Woodbury, New York (1997).

PVDF

Not all piezoelectric materials are crystals or ceramics. A number of natural materials, for example, including wood, bone, collagen, wool, and silk, exhibit weak piezoelectric behavior (3). Polyvinylidene difluoride (PVDF, Figure 1.13), a semicrystalline synthetic polymer composed of doubly fluorinated ethane units, $(CH_2\text{-}CF_2)_n$, exhibits relatively strong piezoelectric characteristics (Table 1.5). Thin sheets of PVDF can be stretched in one direction or in two, perpendicular, directions along the plane of the sheet, then formed into almost any shape. Small piezoelectric charge constants (d_{31} for PVDF is about 10-20% that for PZT ceramics) make PVDF useful for actuators; large voltage constants (g) make it useful for sensors. Further, although the piezoelectric characteristics of PVDF are generally much lower than corresponding characteristics for lead zirconate titanate or barium titanate ceramics, PVDF is unique in that it is not depolarized while subjected to very high alternating electric fields. Consequently, the maximum permissible electric field is one hundred times larger for PVDF than for PZT ceramics. The net result is the maximum strain attained by PVDF can be ten times larger than for PZT. PVDF is a good acoustic impedance match for water or the human body, and a low mechanical quality factor allows a broad resonance bandwidth. The combination of a piezoelectric nature with flexibility and mechanical durability make PVDF attractive for applications including loudspeakers, directional microphones, earphones, and ultrasonic hydrophones.

Irradiated organic films also exhibit piezoelectric properties, and are under investigation for practical applications.

Ceramic elements

Thin Films

Thin films of piezoelectric materials, deposited on various substrates, are a developing approach for actuating miniature components. Zinc oxide (ZnO) is currently the most widely investigated material for thin film applications, but various formulations of lead zirconate titanate materials also show promise, and offer significantly better piezoelectric coefficients. Potential applications for PZT thin films include micro actuators and micro transducers, memory devices, and sensors, including surface acoustic wave (SAW) devices (see **Sensors**). Thin films can be applied to a substrate surface through physical processes (especially sputtering, but also by evaporation or plating) or chemical processes (e.g., sol-gel techniques, vapor deposition).

Photostriction

Photostriction is a phenomenon in which strain is induced in a piezoelectric element solely by illuminating the element. The effect is a consequence of coupling a photovoltaic effect with the piezoelectric properties of the element—illuminating the element generates a voltage sufficient to induce the inverse piezoelectric effect (a piezoelectric element subjected to voltage expands or contracts). Lanthanum-containing lead zirconate titanate (PLZT) ceramics, transparent materials that also are under investigation for their electro-optic properties, exhibit photostriction when irradiated with near-ultraviolet light (4). Using bilaminar elements constructed from PLZT ceramics doped with tungsten (WO_3), investigators have created prototypes of a photo-driven relay, a micro walking device, and a photoacoustic device (photo-induced mechanical resonance at 75 Hz), each of which had neither electric circuits nor leads (4).

Table 1.7 Performance characteristics of APC piezoelectric materials

APC Material	Voltage Limits (V/mil) DC	AC	Output Power (watts/inch^2)	Pressure to Voltage (V) Standard*	Metric**
840	18	9	40	4.35 x ( )	26 x ( ) x 10^{-3}
850	15	8	20	4.36 x ( )	26 x ( ) x 10^{-3}
855	10	5	20	3.27 x ( )	19.5 x ( ) x 10^{-3}
880	20	10	30	4.19 x ( )	25 x ( ) x 10^{-3}

* *volts* $= g_{33}$ • *force (pounds) x ceramic thickness (inches) / surface area (inches^2)*
** *volts* $= g_{33}$ • *force (newtons) x ceramic thickness (meters) / surface area (meters^2)*

Piezoelectric Ceramics from APC

APC International, Ltd. offers an extensive selection of stock and custom-prepared piezoelectric ceramic elements, manufactured from highest purity lead zirconate and lead titanate, with or without secondary materials. Brief general descriptions of APC ceramic materials follow; specific piezoelectric characteristics are summarized in Table 1.6. Performance characteristics of these materials are summarized in Table 1.7; commonly used structural elements are described in Table 1.8.

APC 840, APC 841

Choose APC 840 or APC 841 when high power characteristics are required – generating ultrasonic or high-voltage energy, for sonar devices, etc.:

- ultrasonic cleaners
- ultrasonic atomizers
- ultrasonic micro-bonding apparatus
- underwater echo sounders
- high frequency transducers
- high stress pressure sensors
- squeeze-type gas ignitors
- high power actuators
- vibratory motors
- transformers

Important characteristics of APC 841: a high piezoelectric charge constant (d_{33}), relative to reference values for ceramics of this formulation, produces greater power per volume of material; a high mechanical quality factor reduces mechanical loss and allows a lower operating temperature; a low dissipation factor ensures cooler, more economical operation.

APC 850

Use this material for low-power resonance or non-resonance devices, when high coupling and/or high charge sensitivity are required:

- flow meters
- thickness gauges
- underwater hydrophones
- pressure sensors
- accelerometers
- impact-type gas ignitors
- precise movement control
- microphones
- medical monitoring
- bimorphs

APC 850 is an excellent choice for fluid flow or level sensors (ultrasonic or Doppler flow meters, gas flow sensors, ultrasonic level sensors), or for ultrasonic NDT/NDE applications, for accurate evaluations of automotive, aerospace, or structural products (inspection/evaluation of transmission lines, steel storage tanks, construction materials, structural welds). APC 850 also is an ideal material for high performance transducers for medical imaging applications.

A high dielectric constant, high coupling, high charge sensitivity, high density with a fine grain structure, and a high Curie point are characteristic of APC 850. The material produces a clean, noise-free frequency response.

Table 1.6 Piezoelectric properties of APC lead zirconate titanate materials

Unit	Symbol	APC Ceramic 840	841	850	855	880
Relative Dielectric Constant						
1	K^T	1250	1350	1750	3300	1000
Dielectric Dissipation Factor (Dielectric Loss) *Measured: 1 kHz @ low field*						
%	$\tan \delta$	0.4	0.35	1.4	1.3	0.35
Curie Point*						
°C	T_c	325	320	360	250	310
Electromechanical Coupling Factor						
%	k_p	0.59	0.60	0.63	0.68	0.50
	k_{33}	0.72	0.68	0.72	0.76	0.62
	k_{31}	0.35	0.33	0.36	0.40	0.30
	k_{15}	0.70	0.67	0.68	0.66	0.55
Piezoelectric Charge Constant						
10^{-12} C/N	d_{33}	290	300	400	630	215
or	$-d_{31}$	125	109	175	276	95
10^{-12} m/V	d_{15}	480	450	590	720	330
Piezoelectric Voltage Constant						
10^{-3} Vm/N	g_{33}	26.5	25.5	26	21.0	25
or	$-g_{31}$	11	10.5	12.4	9.0	10
10^{-3} m²/C	g_{15}	38	35	36	27	28
Young's Modulus						
10^{10} N/m²	Y^E_{11}	8	7.6	6.3	5.9	9
	Y^E_{33}	6.8	6.3	5.4	5.1	7.2
Frequency Constants						
Hz•m	N_L**	1524	1700	1500	1390	1725
or m/s	N_T***	2005	2005	2032	2079	2110
	N_P****	2130	2055	1980	1920	2120
Density						
10^3 kg/m³	ρ	7.6	7.6	7.7	7.7	7.6
Mechanical Quality Factor						
1	Q_m	500	1400	80	65	1000

All values nominal; measurements made 24 hours after polarization.
* *Maximum operating temperature = Curie point / 2.*
** N_L = *longitudinal mode*
*** N_T = *thickness mode*
**** N_P = *radial mode*

Comparative values for quartz:
relative dielectric constant (K): 5
electromechanical coupling factor (k_{33}): 0.09
piezoelectric charge constant (d_{33}): 2.3
piezoelectric voltage constant (g_{33}): 57.8
mechanical quality factor (Q_m): >10^5

APC 855

These materials are suited to applications similar to those in which APC 850 is used, but when high permittivity and high piezoelectric charge coefficients are required.

Although its properties are somewhat different from those of APC 850, APC 855, like APC 850, is an excellent material for fluid flow or level sensors, ultrasonic NDT/NDE applications, and high performance transducers for medical imaging applications.

APC 880

Use this ceramic when the highest electrical drive is required. High dielectric stability and low mechanical loss under high drive conditions make APC 880 an excellent choice for:

- ultrasonic welding
- ultrasonic mixing / dispersion
- ultrasonic surgery
- cryogenic SEMs

Suggested Reading

Cady, W.G. *Piezoelectricity*
McGraw-Hill, New York (1946);
reprint: Dover Press, New York (1964).

Jaffe, B., W. Cook, and H. Jaffe *Piezoelectric Ceramics*
Academic Press, London (1971).

Mason, W.P. *Applications of Acoustical Phenomena*
Jour. Acoustical Soc. Amer., 50 (5), part 2 (1980).

Mason, W.P. *Piezoelectricity, Its History and Applications*
Jour. Acoustical Soc. Amer., 70 (6) (1981).

American National Standard on Piezoelectricity
ANSI / IEEE Standard 176 (1987).

Piezoelectric Ceramic for Sonar Transducers (Hydrophones and Projectors)
Military Standard DOD-STD-1376A(SH) (1984).

References

1. Tressler, J.F. and K. Uchino, *Piezoelectric Composite Sensors*
Vol. II/Appendix 37 in *Acoustic Transduction — Materials and Devices* Annual Report (1 January 1999 to 31 December 1999) Office of Naval Research, Contract No: N00014-96-1-1173 (June 2000).

2. Park, S.-E., S. Wada, L.E. Cross, and T.R. Shrout, *Crystallographically Engineered BaTiO$_3$ Single Crystals for High-Performance Piezoelectrics*
Vol. II/Appendix 29 in *Acoustic Transduction — Materials and Devices* Annual Report (1 January 1999 to 31 December 1999) Office of Naval Research, Contract No: N00014-96-1-1173 (June 2000).

3. Silk, M.G., *Ultrasonic Transducers for Nondestructive Testing* (1984). Adam Hilger Ltd, Bristol.

4. Uchino, K. and P. Poosanaas, *Photostriction in PLZT and its Applications*
Vol. IV/Appendix 54 in *Acoustic Transduction — Materials and Devices* Annual Report (1 January 1998 to 31 December 1998) Office of Naval Research, Contract No: N00014-96-1-1173 (Apr. 1999).

Request annual reports to Office of Naval Research from:
Office of Naval Research
Regional Office Chicago
536 S. Clark Street
Room 208
Chicago, IL 60605-1588.

Table 1.8   Modes of vibration for common piezoelectric ceramic shapes

Axis	Polarization Direction	Applied Field: Voltage Output	Mode of Vibration: Displacement
Plate			
			length or transverse (l or w) thickness (h)
Disc			
			radial (r) thickness (h)
Ring			
			radial (r) thickness (h)
Bar			
			length (l)
Rod			

Frequency Constant	Capacitance	Static Displacement	Static Voltage
$N_L = f_r l$ $N_L = f_r w$ $N_T = f_r h$	$C_S = \dfrac{K^T_{33}\varepsilon_0 lw}{h}$	$\Delta l = \dfrac{d_{31}Vl}{h}$ $\Delta w = \dfrac{d_{31}Vw}{h}$ $\Delta h = d_{33}V$	$V = \dfrac{g_{31}F_1}{l}$ $V = \dfrac{g_{31}F_2}{w}$ $V = \dfrac{g_{33}F_3 h}{lw}$
$N_p = 2f_r r$ $N_T = f_r h$	$C_S = \dfrac{K^T_{33}\varepsilon_0 \pi r^2}{h}$	$\Delta r = \dfrac{2d_{31}Vr}{h}$ $\Delta h = d_{33}V$	$V = \dfrac{g_{31}F_r}{2\pi r}$ $V = \dfrac{g_{33}F_3 h}{\pi r^2}$
$N_{ring} = f_r(OD - ID)$ $N_T = f_r h$	$C_S = \dfrac{K^T_{33}\varepsilon_0 \pi(OD^2 - ID^2)}{4h}$	$\Delta r = \dfrac{d_{31}V(OD - ID)}{2h}$ $\Delta h = d_{33}V$	$V = \dfrac{g_{31}F_r}{2\pi(OD - ID)}$ $V = \dfrac{4g_{33}F_3 h}{\pi(OD^2 - ID^2)}$
$N_{axial} = f_r l$	$C_S = \dfrac{K^T_{33}\varepsilon_0 wh}{l}$	$\Delta l = d_{33}V$	$V = \dfrac{g_{33}F_3 l}{wh}$ $V = \dfrac{g_{31}F_1}{w}$ $V = \dfrac{g_{31}F_2}{h}$
$N_{axial} = f_r l$	$C_S = \dfrac{K^T_{33}\varepsilon_0 \pi r^2}{4l}$	$\Delta l = d_{33}V$	$V = \dfrac{g_{33}F_3 l}{\pi r^2}$ $V = \dfrac{g_{31}F_r}{2\pi r}$

Table 1.8 Modes of vibration for common piezoelectric ceramic shapes *(cont'd)*

Axis	Polarization Direction	Applied Field: Voltage Output	Mode of Vibration: Displacement
Cylinder End Electrode			
			length (l) radial (r)
Cylinder Wall Electrode			
			length (l) radial (r)
Plate Bender			

Cylinder with Striped Electrodes

Disk Bender

Frequency Constant	Capacitance	Static Displacement	Static Voltage
$N_L = f_r l$ $N_{ring} = f_r(OD - ID)$	$C_S = \dfrac{K^T_{11}\varepsilon_0\pi(OD^2 - ID^2)}{4l}$	$\Delta l = d_{33}V$ $\Delta r = \dfrac{d_{31}V(OD - ID)}{2l}$	$V = \dfrac{4g_{33}F_3 l}{\pi(OD^2 - ID^2)}$ $V = \dfrac{g_{31}F_r}{2\pi(OD - ID)}$
$N_{axial} = f_r l$ $N_T = f_r(OD - ID)/2$	$C_S = \dfrac{2K^T_{33}\varepsilon_0\pi l}{\ln(OD/ID)}$	$\Delta l = \dfrac{2d_{31}Vl}{(OD - ID)}$ $\Delta r = d_{33}V$	$V = \dfrac{g_{31}F_1(OD - ID)}{l(OD - ID)}$ $V = \dfrac{g_{33}F_3(OD - ID)}{2(OD^2 - ID^2)}$
$N_L = \dfrac{3f_r l^2}{h}$ $N_L = \dfrac{3f_r l^2}{h}$	$C_S = \dfrac{K^T_{33}\varepsilon_0 lw}{h}$	$\Delta h = \dfrac{3d_{31}Vl^2}{2h^2}$ $\Delta h = \dfrac{3d_{31}Vl^2}{h^2}$	$V = \dfrac{3g_{31}F_3 l}{2wh}$ $V = \dfrac{3g_{31}F_3 l}{4wh}$

Hemisphere
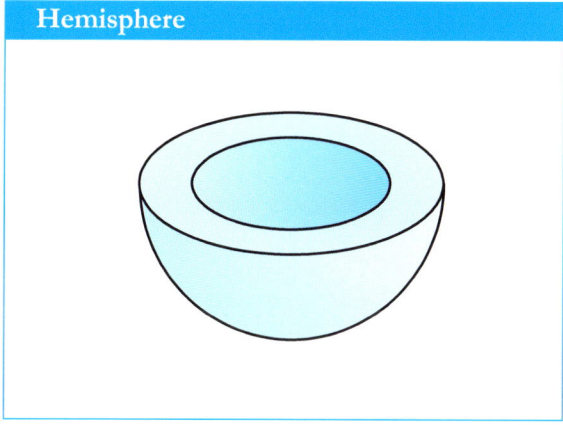

General

Equations valid for:
(A) plate, disc & ring where r, l, & w>>h
(B) bar where l>>r, w, & h (5 to 10X)
(C) cylinder & rod where h>>r (5 to 10X)
(D) plate bender equations are for fully electroded cantilever mount. For beam mount multiply f_r by 2.8, Δt by 0.25, & V by 0.25
Constants g_{31} & d_{31} are negative values resulting in negative strain & negative voltage. All variables are in metric (MKS) units. Each material has voltage, stress, and temperature limitations.

generators

Piezoelectric ceramic elements can generate voltages sufficient to spark across an electrode gap, and consequently are used as ignitors in fuel lighters, gas stoves, welding equipment, and other such apparatus. Piezoelectric fuel ignition devices are smaller, simpler, and less expensive than systems incorporating permanent magnets or high voltage transformers and capacitors.

The volume of the ceramic element and the amount of stress exerted on the element are key factors in converting mechanical input to electrical energy. The stress on the element is the ratio of the applied force to the surface area of the element. Consequently, when the composition of the ceramic, the volume of the ceramic element, and the applied force are constant, the element that has the smallest surface area will generate the most electrical energy.

In *squeeze-type* piezoelectric fuel ignitors a static mechanical energy input—very low frequency, relative to the resonance frequency of the ceramic—generates the electrical energy for ignition. In the *impact ignition* design a spring-loaded hammer delivers a dynamic input to the ceramic element. The pressure wave generated by the hammer striking the element once is reflected multiple times in both the element and the hammer. Until the flashover at the spark gap, stress varies along the height of the ceramic element, and accurate values for voltage must be calculated by integration over the height of the element. More easily calculated approximate values usually are sufficient for designing these simple devices.

In flashbulbs, fuses for detonating explosives, or other disposable items, the piezoelectric generator is intentionally depolarized. The goal in these one-use applications is to extract from the generator as much electrical energy as possible. An open circuit system will maximize energy density; very high voltages can be reduced by incorporating a high capacitance device into the system, in parallel with the generator.

Alternatively, the electrical energy generated by a piezoelectric element can be stored. Techniques used to manufacture multilayer capacitors have been used to construct multilayer piezoelectric generators. Because piezoelectric batteries do not exhibit electromagnetic interference—an undesirable attribute of electrochemical batteries—high charge / low voltage piezoelectric batteries are ideal power sources for sensitive electronic circuits.

Generators

A piezoelectric ceramic element will generate electrical energy from a mechanical energy input. In a properly designed generator, voltage will increase almost linearly with increasing stress. Typically, the generated voltage is made to spark across an electrode gap (Figure 2.1a), to ignite a combustible gas. In these fuel ignition applications, a piezoelectric system is much simpler—and much more compact—than alternative systems incorporating permanent magnets or high voltage transformers and capacitors. Alternatively, the electrical energy can be stored and used as a power source (Figure 2.1b). Piezoelectric batteries do not exhibit electromagnetic interference, a characteristic of electrochemical batteries. Consequently, high charge / low voltage piezoelectric batteries are ideal for powering interference-sensitive electronic circuits.

In a piezoelectric generator, charge generation is directly related to the extent to which the ceramic element is deformed. Consequently, the configuration of the element and the manner in which it is mounted are important contributors to the performance of the generator. A tall ceramic cylinder, constrained at its ends, will radially expand much more readily than will a short disk of equal volume, under similar constraint, and thus will convert significantly more of the mechanical energy input into electrical energy. The electrical energy must be rapidly dissipated from the generator, or the electric field could partially or completely depolarize the ceramic element.

Static Input

Open Circuit System

Static or near-static charges (input frequencies very low, relative to the resonance frequency of the ceramic) can depolarize a generator more rapidly than dynamic charges of the same magnitude. In an open circuit system, a piezoelectric ceramic generator will not be depolarized if the electric field induced by the stress has the same direction as the field in which the ceramic element was polarized.

Stress applied to a ceramic element parallel to the direction of polarization will deform the element across the direction of polarization and create an electric field with the same polarity as the field used to polarize the element. For an open circuit system, the voltage generated in this manner can be calculated from Equation 2.1. Note that a compressive stress (negative sign) generates positive voltage.

Figure 2.1   Electrical energy generation by a piezoelectric ceramic element

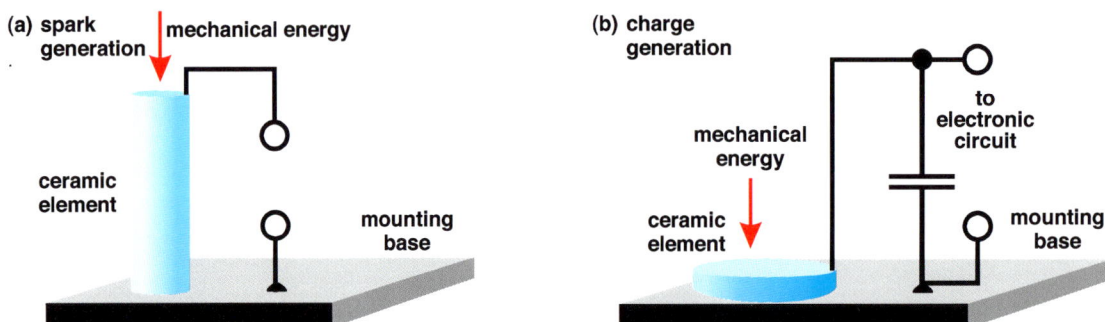

(a) spark generation   mechanical energy
ceramic element
mounting base

(b) charge generation
mechanical energy
ceramic element
to electronic circuit
mounting base

Equation 2.1

$$V = - (g_{33}hT)$$

where

V = voltage

g_{33} = piezoelectric voltage constant*

h = height (thickness) of ceramic element

T = stress on element**

* electric field generated per unit of mechanical stress applied

** applied force / surface area of ceramic element, in meters^2

Some of the mechanical energy applied to the ceramic element will be expended in deforming the element. Assuming no other losses, the remainder of the input energy will be converted to useable electrical energy:

Equation 2.2

$$W_t = W_d + W_e$$

where

W_t = total mechanical energy input

W_d = energy expended to deform ceramic element

W_e = energy in electric field of ceramic element

W_t, W_d, or W_e can be quantified from Equation 2.3, 2.4, or 2.5, respectively. Note that the volume of the ceramic element and the amount of stress exerted on the element are key factors in the conversion of mechanical input to electrical energy.

The stress on the element, T, is the ratio of the applied force to the surface area of the element. Consequently, when the composition of the ceramic, the volume of the ceramic element, and the applied force are constant, the element that has the smallest surface area will generate the most electrical energy.

Equation 2.3

$$W_t = [(vol) (s^D_{33}T^2)] / 2$$

Equation 2.4

$$W_d = [(vol) (1 - k_{33}^2) (s^D_{33}T^2)] / 2$$

Equation 2.5

$$W_e = [(vol) (k_{33}^2) (s^D_{33}T^2)] / 2$$

where

vol = volume of ceramic element

s^D_{33} = elastic compliance*

T = stress on element

k_{33} = coupling factor**

* strain produced per unit of stress applied to ceramic element; constant electric displacement

** effectiveness of mechanical energy / electrical energy conversion

Alternatively, W_e can be determined from electrical characteristics of the system:

Equation 2.6

$$W_e = [(1 - k_{33}^2) (C_0V^2)] / 2$$

where

k_{33} = coupling factor

C_0 = capacitance of ceramic element (input frequency << resonance frequency)

V = voltage

Very high voltages are characteristic of an unmodified system. To attain a practical voltage, a capacitor, C_p, with a much larger capacitance than that of the piezoelectric generator, can be incorporated into the system in parallel with the generator (Figure 2.2). Voltage will be reduced linearly with the increase in total capacitance. Further, the capacitor can power an electronic circuit (see **Solid State Batteries**).

Figure 2.2 Parallel capacitor reduces voltage and stores electrical energy

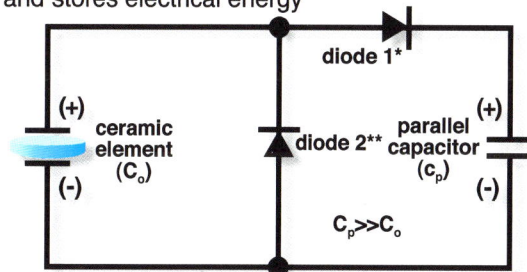

* prevents charge from flowing back to generator when compressive force is removed

** protects diode 1 from high reverse voltages (e.g., ceramic element incompletely depolarized)

Because available energy is related to the square of the voltage, capacitor C_p will greatly reduce available energy in addition to reducing the voltage. There are two approaches to minimizing the energy loss: adjusting the capacitance by constructing the generator from multiple thin layers of ceramic or adjusting the impedance by incorporating a transformer into the system. In the first approach, the generator is constructed from multiple very thin layers of ceramic (1/10 mm or less), alternated with electrodes, rather than from a single, much thicker ceramic element. Capacitance C_0 will be higher for a generator constructed in this manner and, because the multiple-layer construction creates a large surface area to volume ratio, the generator will generate a high charge and comparatively low voltage:

Equation 2.7

$$V = (h/n)\, g_{33}\, T$$

where

V = voltage
h = height (thickness) of ceramic element
n = number of ceramic layers
g_{33} = piezoelectric voltage constant
T = stress on element

Because C_0 is higher and V is lower, relative to corresponding values for a single thick ceramic element, the parallel capacitance C_p incorporated into the system can be smaller.

Alternatively, if a larger and more complicated system is acceptable, a transformer can be used to adjust the impedance (Figure 2.3).

Short-Circuited System

If a short-circuited piezoelectric ceramic element is stressed, electric charge will be generated by both linear processes (deformation of domains) and nonlinear processes (reversible or irreversible displacements of domain walls). The electric displacement produced by the linear processes is:

Equation 2.8

$$Q / A = D = d_{33}\, T$$

where

Q = total generated charge
A = surface area of ceramic element
D = electric displacement
d_{33} = piezoelectric charge constant*
T = stress on element
* electric polarization generated per unit of mechanical stress applied

Because a significant nonlinear component is added to the linear piezoelectric effect, measured electric displacements agree with calculated values only at very low voltages. The additional, nonlinear contribution can cause displacement values to rise more sharply with increasing stress than would be expected from calculations that assume a linear input alone. Irreversible displacements of domain walls can depolarize the ceramic element, and consequently are acceptable only if the generator is intended for one-time use. Depolarization in a generator that is intended to function repeatedly can be ascertained by measuring the charge developed during successive charge / discharge cycles. Equivalent curves indicate there has been no depolarization.

Repeated-Use Applications:
Generating Electrical Energy without
Depolarizing the Generator

In piezoelectric fuel ignition systems in cigarette lighters, gas ovens, or other repeated-use applications, the generator must not be depolarized by the electric field it creates. In these applications, the ceramic element will act like an electrically open system until the flashover at the spark gap, but at that time the resistance of the conducting spark gap will put the system under load. In these systems, total available electrical energy, W_{ee}, can be quantified from Equation 2.9 or 2.10; energy density, w_{ee}, can be determined from Equation 2.11:

Equation 2.9

$$W_{ee} = [(\varepsilon^T_{33}) (A / h) (V^2)] / 2$$

Equation 2.10

$$W_{ee} = (C_0 V^2) / 2$$

Equation 2.11

$$w_{ee} = (\varepsilon^T_{33} g_{33}{}^2 T^2) / 2$$

where

ε^T_{33} = permittivity at constant stress

A = surface area of ceramic element

h = height (thickness) of ceramic element

V = voltage

C_0 = capacitance of ceramic element

g_{33} = piezoelectric voltage constant

T = stress on element

One-Time Use Applications: Depolarizing the Generator

Applications in which a piezoelectric generator is intentionally depolarized usually involve disposable items, such as flashbulbs or fuses for detonating explosives. The goal in these applications is to extract from the generator as much electrical energy as possible. An open circuit system will maximize energy density, but flashover and dielectric breakdown make this energy available only for a very short time. Very high voltages can be reduced by incorporating a high capacitance device into the system, in parallel with the generator. The capacitor will reduce the voltage linearly, in direct correlation with the increase in the total capacitance of the system (see **Open Circuit System** above).

Dynamic Input (Impact Ignition)

Fuel-igniting piezoelectric devices in which a static mechanical energy input generates electrical energy are called squeeze-type ignitors. Alternatively, some piezoelectric ignitor systems employ an impact ignition design that delivers a dynamic input. A spring-loaded hammer strikes the ceramic element once, but the pressure wave generated by the impact is reflected multiple times in both the ceramic element and the hammer. The exact behavior of the wave depends on the elasticity of the hammer and the acoustical properties of the ceramic and the hammer. Until the flashover

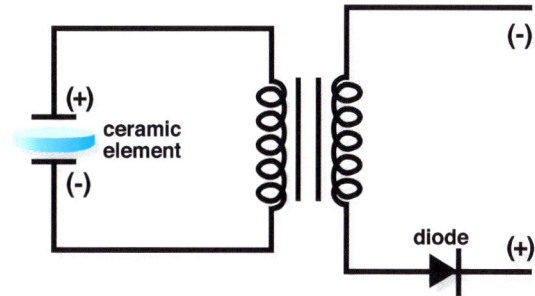

Figure 2.3 A transformer is an alternative for regulating electrical energy

at the spark gap, the generator acts like an open circuit system, as it does in a static input device, but stress varies along the height of the ceramic element. If an exact value for voltage is critical, V must be calculated by integration over the height of the element. For a simple igniting device, however, Equation 2.1 provides a practical approximation of V. Similarly, peak voltage can be measured, then used in Equation 2.10 to give an approximate, but conservative, estimate of the available energy. Nonlinear phenomena usually will supplement this value but, again, for these simple devices more accurate calculations usually are unnecessary.

Solid State Batteries

Because they do not create electromagnetic interference, multilayer piezoelectric generators are excellent solid state batteries for electronic circuits. A piezoelectric generator constructed from sub-millimeter-thick ceramic layers alternated with electrodes will exhibit a greater capacitance to voltage ratio than a generator constructed from a single thick ceramic element, so the parallel capacitance needed to reduce the voltage will be smaller (Equation 2.7). To maximize the energy available from the parallel capacitor, the capacitances of the generator and the parallel capacitor should be equal ($C_0 = C_p$).

sensors

A sensor converts a physical, chemical, or other input, such as acceleration or pressure, into an electrical signal, and routes the signal to a data processing system. Often, the signal elicits a response from the system–the seat belts in an automobile are locked when a sensor detects rapid deceleration, for example. *Transducers*–devices that produce and receive audible sound or ultra-sound signals–also employ piezoelectric ceramic elements in a sensory capacity (to measure distances, etc.). These devices are discussed in Chapter 5.

There are two basic constructions of piezoelectric sensors: *axial sensors* and *flexional sensors*. An axial sensor employs a rigid piezoelectric ceramic element to sense a force exerted parallel to the direction in which the element is polarized. Common applications for axial sensors include monitoring acceleration and detecting automobile engine knock.

In the most common design of flexional sensor, two plates or strips of piezoelectric ceramic, polarized in the third (thickness) direction, are bonded together to produce a flexible bilaminar element. If the two plates are joined with the direction of polarization opposing, and electrodes are applied to the outer surfaces, the plates are in series. Joining the plates with polarization in the same direction produces a parallel element. In typical applications, one end of the flexible element is fixed and the input to be measured acts on the free end (cantilever mounting). A force exerted in the thickness direction at the free end of the element causes the sensor to bend away from the plane perpendicular to the direction of polarization, creating tension in the ceramic layer to which the force is applied and compression in the opposing layer, and producing a voltage in each layer. In a series element, the electric fields and voltages produced by the tensile and compressive forces have the same

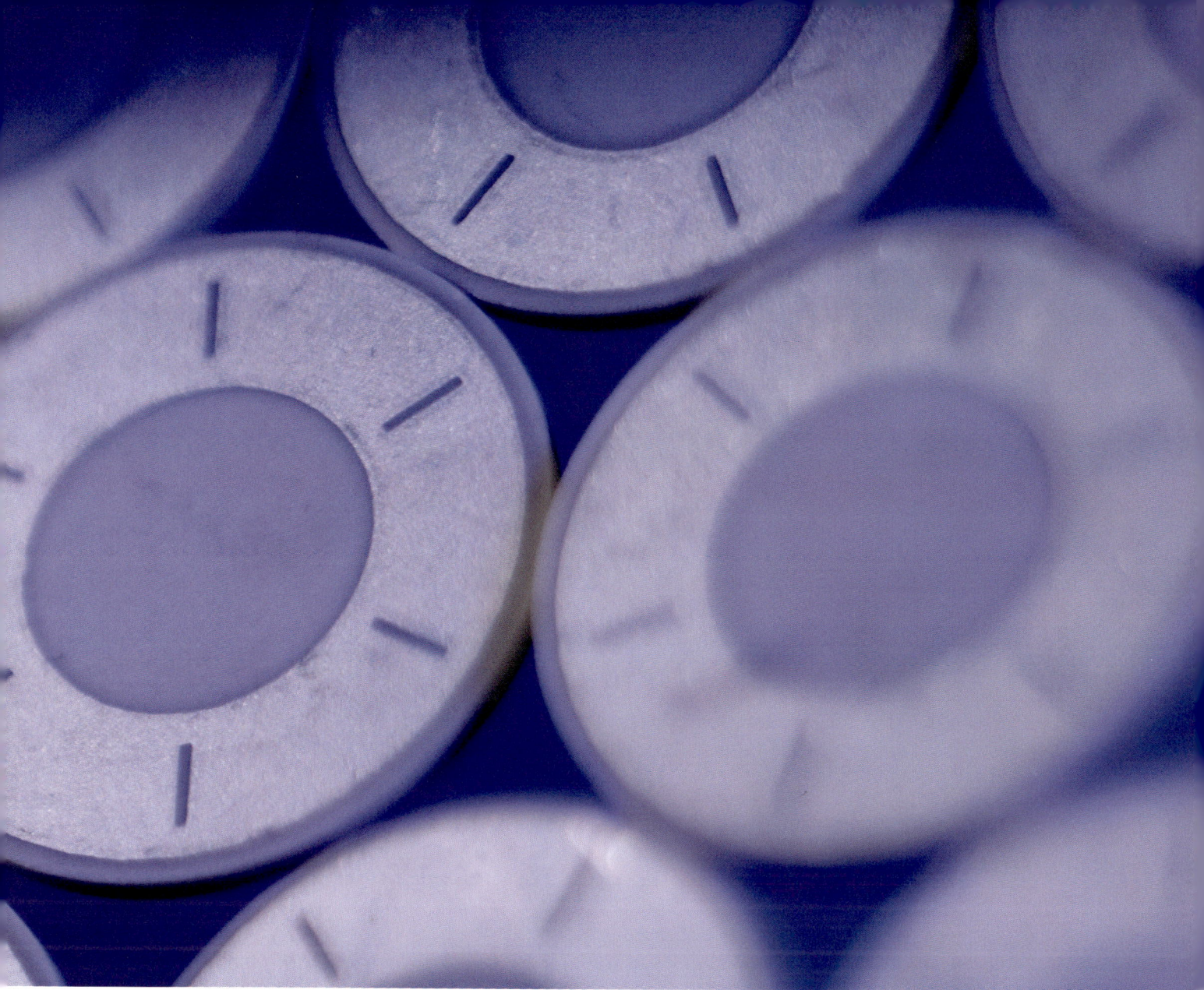

direction. In a parallel element, the electric fields and voltages generated in the two plates are opposing.

Relative to axial sensors, flexional sensors have much lower stiffness, lower mechanical impedance, and lower electrical impedance. These characteristics make flexional sensors better adapted to sensing subtle mechanical input, relative to axial sensors, and compatible with relatively simple amplifiers. For long-term monitoring applications, single crystals are superior to polycrystalline ceramic elements, because the piezoelectric properties of the former are far more stable. On the other hand, polycrystalline ceramics make superior force and displacement sensors, because they exhibit high mechanical strength and better resistance to adverse temperatures, pressures, and humidity levels.

A typical piezoelectric sensor will generate a signal only when it experiences a change in the applied force or pressure—under a static input, free charge carriers in the ceramic element neutralize the charges on the dipoles, effectively discharging the element.

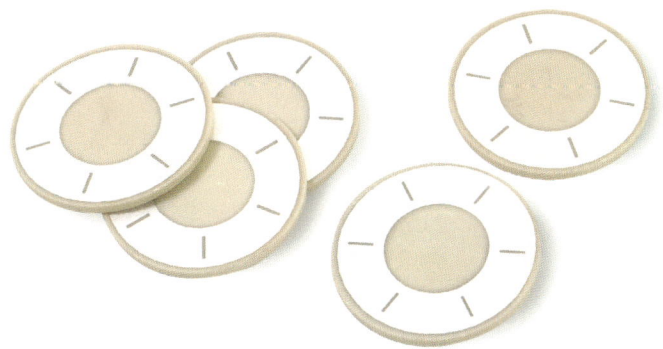

Sensors

Sensors are intermediaries between the physical world and systems that electronically process and store information: a sensor accepts a particular physical, chemical, or biological input, converts the information into an electrical signal, and delivers the signal to a data processing system. Often, the system incorporating the sensor exhibits a response in reply to the input (e.g., the seat belts in an automobile are locked when a sensor detects rapid deceleration; an indicator light is lit when a fluid level drops below specification).

Figure 3.1 shows the two basic constructions of piezoelectric sensors: axial sensors and flexional sensors. An axial sensor senses a force exerted parallel to the direction in which the piezoelectric ceramic element is polarized, and generates an electrical energy signal in the same direction (Figure 3.2a). Because both the mechanical strain and the electrical signal are expressed in direction 3, axial sensors are called d_{33} sensors or 33-mode sensors. A flexional (bending) sensor also measures a force exerted in the direction of polarization of the ceramic element, but the force causes the ceramic element or elements to bend (elongate or contract) along the plane perpendicular to the direction of polarization (Figure 3.2b). Consequently, flexional sensors are called d_{31} sensors or 31-mode sensors.

Both axial sensors and flexional sensors serve in the signal-receiving roles discussed here. In addition, however, transducers—devices that produce and receive audible sound or ultrasound signals (to locate objects, measure distances, measure flow velocity, etc.)—also employ piezoelectric ceramic elements in a sensory capacity. In these applications a single ceramic element can be used both to transmit the output signal and receive (sense) the reflected signal, or separate elements can be used in the transmission and reception (sensory) roles. These devices are discussed in **Transducers**.

It is important to recognize that a typical piezoelectric sensor will generate a signal only when it experiences a change in the applied force or pressure. Under a static input, free charge carriers in the ceramic element migrate toward the dipoles, neutralizing the charges on the dipoles and thus effectively electrically discharging the element. The charge also will drain across the input resistance of the device used to measure the signal from the sensor. A stress upsets the

Figure 3.1 Basic construction of piezoelectric sensors

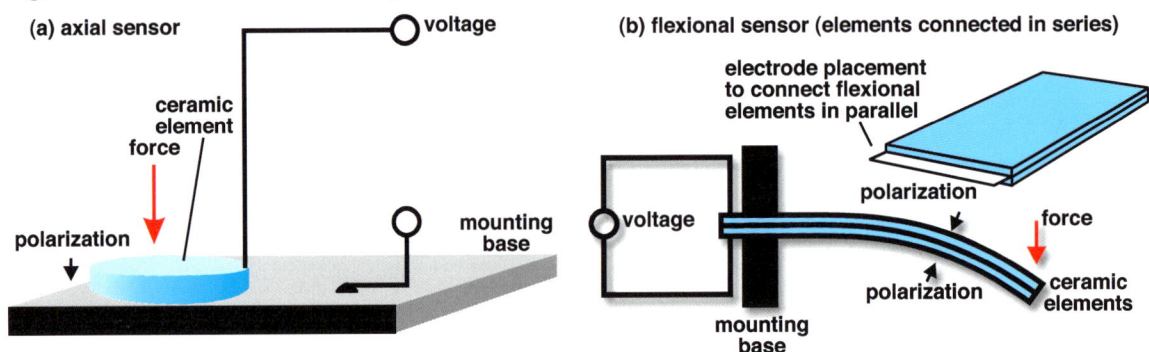

(a) axial sensor

voltage
ceramic element
force
polarization
mounting base

(b) flexional sensor (elements connected in series)

electrode placement to connect flexional elements in parallel
polarization
voltage
force
polarization
ceramic elements
mounting base

Figure 3.2 Relationship between mechanical input and electrical signal output by a sensor

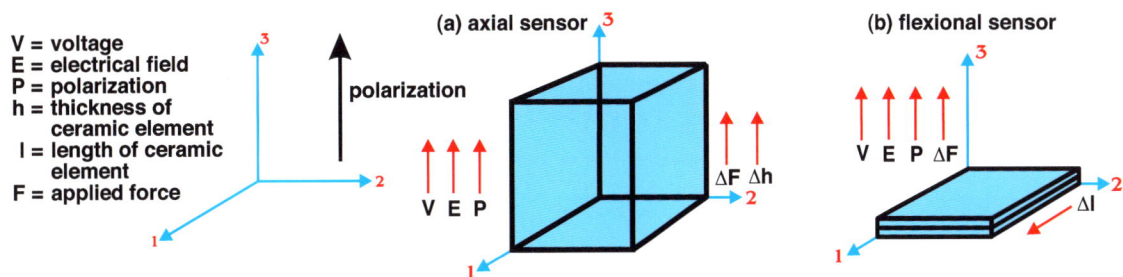

V = voltage
E = electrical field
P = polarization
h = thickness of
   ceramic element
l = length of ceramic
   element
F = applied force

polarization

(a) axial sensor

ΔF Δh

V E P

(b) flexional sensor

V E P ΔF

Δl

balanced state and restores an electric charge, but if the stress is maintained the charge will drain again. In practice, systems for measuring low-frequency signals—input frequencies far below the resonance frequency of the system—are conveniently described by the time constant. The time constant of the system is the product of the capacitance of the ceramic element, C_0, and the input resistance of the electronic circuit. As a rule of thumb, the time constant should be at least ten times the period time of the input signal. For example, to measure a signal with a frequency of 10 Hz, the time constant must be at least 1s. There are three alternatives that keep the input resistance acceptably low, while enabling low frequency inputs to be measured: constructing the sensor from multiple parallel-connected layers, incorporating a charge amplifier in the system, or incorporating a capacitor in the system, in parallel with the sensor. The latter approach will reduce the signal, however.

For sensors intended for long-term monitoring applications, single crystals, oriented and cut along specific directions, are superior to polarized polycrystalline ceramics, because the piezoelectric properties of single crystals are far more stable. On the other hand, polycrystalline ceramics make superior force and displacement sensors, because they exhibit high mechanical strength and better resistance to adverse environmental conditions (temperature, pressure, humidity). The properties of ceramics are relatively easily reproduced, and these materials can be manufactured in a relatively unlimited range of physical shapes and dimensions.

Axial Sensors

The charge, Q, developed by applying a force to a piezoelectric sensor is a product of the capacitance of the piezoelectric ceramic element and the voltage generated as a consequence of applying the force. Charge is independent of the dimensions of the ceramic element:

Equation 3.1

$Q = C_0 V$

where

C_0 = capacitance of ceramic element
V = voltage

or, for an axial sensor:

Equation 3.2

$Q = -(d_{33}F)$

where

d_{33} = piezoelectric charge constant*
F = applied force
* polarization generated per unit of
  mechanical stress applied

If multiple elements are stacked together and connected in parallel, however, the force will act simultaneously on each element (Figure 3.3), and the charge will be:

Equation 3.3

$Q = -(d_{33}F)n$

where

n = number of ceramic elements

In contrast, the voltage, V, developed

Figure 3.3 Multiple element axial sensor: generated charge is proportional to number of piezoelectric elements

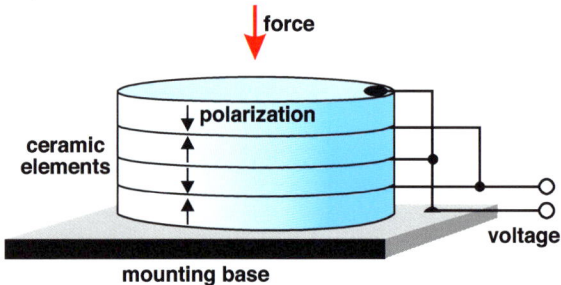

by applying a force to a piezoelectric sensor is directly related to the height and inversely related to the cross-sectional area of the ceramic element. The open circuit voltage for a sensor consisting of either a single ceramic element or multiple elements connected in parallel is:

Equation 3.4

$$V = -(g_{33}hT)$$

where

g_{33} = piezoelectric voltage constant*

h = height (thickness) of ceramic element

T = stress on element**

* electric field generated per unit of mechanical stress applied

** applied force / surface area of element

Whether a sensor is constructed from a single ceramic element or from multiple elements connected in parallel, voltage will be the same, and will increase almost linearly with increasing stress. Greater charge and higher capacitance, and thus lower impedance, make the multiple element arrangement more appealing, just as multilayer generators offer these advantages relative to single element generators (see **Generators**).

Basic applications for axial sensors include monitoring acceleration and detecting automobile engine knock. The simplist acceleration sensor is made by securing a disk of piezoelectric ceramic between a mounting base and an inactive seismic mass (Figure 3.4a). During acceleration in the direction in which the ceramic is polarized the seismic mass exerts a force on the ceramic element, and the element generates an electrical signal proportional to the rate of acceleration. A high-density metal, such as osmium or tungsten, generally is used for the seismic mass. In some acceleration sensors the ceramic element itself assumes the role of the seismic mass (Figure 3.4b).

The force exerted by the seismic mass is expressed by:

Equation 3.5

$$F = Ma$$

where

F = force exerted on ceramic element

M = seismic mass

a = acceleration

The stress on the ceramic element is determined by the surface area of the element, as well as the acceleration and the seismic mass:

Figure 3.4 Acceleration sensors

(a) acceleration sensor incorporating piezoelectric ceramic element and separate seismic mass

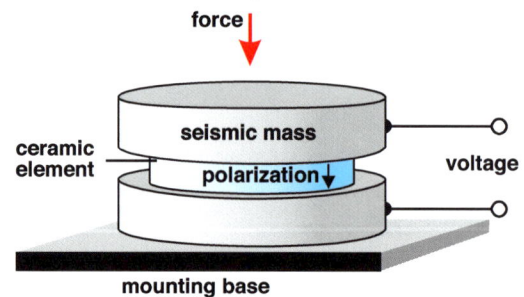

(b) active knock sensor: ceramic element also has seismic mass

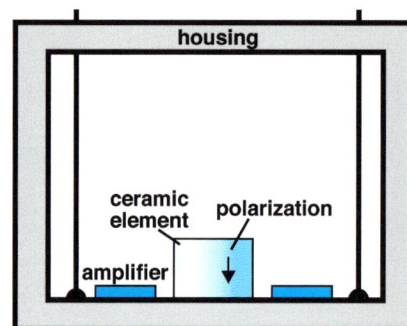

Equation 3.6

$$T = -(M / A)a$$

where

T = stress on ceramic element
M = seismic mass
A = surface area of ceramic element
a = acceleration

An axial acceleration sensor is electrically equivalent to the system depicted in Figure 3.5. For static or near-static conditions (input frequencies much lower than the system resonance frequency), the inductance equivalent to the seismic mass, L_m, can be neglected, which makes the capacitance of the sensor, C_S, the sum of the static capacitance of the ceramic element, C_0, and the "mechanical" capacitance (spring constant), C_m. The load on the sensor is the input impedance of the amplifier, Z_E. To maintain open circuit voltage the capacitance of the sensor, C_S, must equal or exceed the capacitance of the amplifier, C_E. Z_E must equal or exceed $1 / \omega(C_S + C_E)$ ($\omega = 2\pi$ x frequency) and should approximate $1 / \omega C_S$.

The sensitivity, S_V or S_Q, of an acceleration sensor is a function of the electrical energy generated by the sensor and the acceleration responsible for creating that electrical energy. In an open circuit system, S_V is determined by the output voltage, V, and the acceleration, a. In a short-circuited system, S_Q is determined by the generated charge, Q, and the acceleration:

Equation 3.7 (open circuit)

$$S_V = V / a$$

Equation 3.8 (short-circuit)

$$S_Q = Q / a$$

Sensitivity also depends on frequency. Figure 3.6 shows the relationship. S_V and S_Q typically are measured at the flat (frequency-independent) segment of the curve. The upper limit (f_u) of this part of the curve typically is one-half the first resonance frequency, or minimal impedance frequency, f_m, of the sensor. f_m can be determined from:

Figure 3.5   Electrical circuit equivalent to an axial acceleration sensor

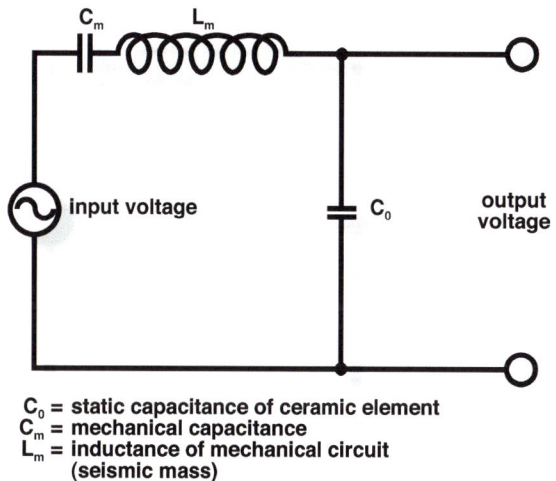

C_0 = static capacitance of ceramic element
C_m = mechanical capacitance
L_m = inductance of mechanical circuit (seismic mass)

Figure 3.6   Sensitivity of an acceleration sensor is a function of frequency

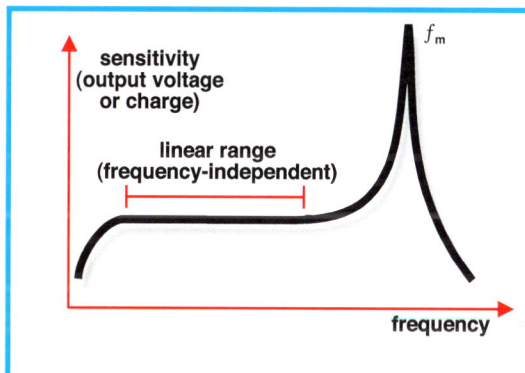

Equation 3.9

$$f_m = (1 / 2\pi) (\sqrt{[K / (M + m / 3)]})$$

where

K = stiffness of ceramic element*
M = seismic mass
m = mass of ceramic element
* applied force / change in height (thickness) of element, or: surface area of element / height of element x elastic compliance (elastic compliance = s^E for short-circuit system, s^D for open circuit system)

The lower limit of the frequency-independent segment of the curve, f_1, depends on Z_E, C_S, and C_E:

Equation 3.10

$$f_1 = 1 / [(2\pi Z_E)(C_S + C_E)]$$

where

Z_E = external (amplifier) impedance
C_S = capacitance of sensor*
C_E = external (amplifier) capacitance
* See Table 1.8.

If the ratio of the seismic mass to the mass of the ceramic element is large (M >> m), the T term in Equation 3.4 for voltage becomes an expression of the acceleration, the seismic mass, and the surface area of the element:

Equation 3.11

$$V = g_{33}h(aM / A)$$

where

V = voltage
g_{33} = piezoelectric voltage constant
h = height (thickness) of ceramic element
a = acceleration
M = seismic mass
A = surface area of element

Note that the acceleration force also cancels the negative sign in Equation 3.4.

As a result, sensitivity for an open circuit system (Equation 3.7; $S_V = V / a$) becomes:

Equation 3.12

$$S_V = g_{33}(h / A)M$$

If the ratio of the seismic mass to the mass of the ceramic element is not large, the value (M + m / 2) replaces M in equations 3.11 and 3.12. If the ceramic element also has the role of the seismic mass (i.e., M = 0), the value (m / 2) replaces M in equations 3.11 and 3.12, and Equation 3.12 can be expressed as:

Equation 3.13

$$S_V = g_{33}(h^2)\rho / 2$$

where

S_V = sensitivity
g_{33} = piezoelectric voltage constant
h = height (thickness) of ceramic element
ρ = density of ceramic material

In practice a small, but finite, external capacitance usually will reduce sensitivity to a value somewhat lower than the theoretical value calculated from Equation 3.12 or Equation 3.13:

Equation 3.14

$$S_P = (S_V)\{1 / [1 + (C_E / C_s)]\}$$

where

S_P = sensitivity (practical)
S_V = sensitivity (theoretical)
C_S = capacitance of sensor
C_E = external capacitance

In practical applications, accelerometers of the simple compression design exhibit shortcomings in stability and directional selectivity. Some of these flaws are overcome by an alternative 33-mode design: a ring-shaped piezo element mounted around a cylindrical central post attached to the mounting base (Figure 3.7). This simple, robust, and stable construction makes center-mounted compression devices well suited for high-level or shock measurements. An inherent source of error that is not addressed by the center-mounted compression design, nor by the simple compression design, is the fact that a temperature increase will create pyroelectric charges, disturbing the measurement. The pyroelectric charges are the electric effect of the strain associated with temperature changes in polar materials. In an axial accelerometer, the charges generated will be associated with the 33 and 31 modes. Although these may have opposite polarity, generally they will not cancel out. Another complicating factor is the temperature dependence of the piezoelectric charge constant, piezoelectric voltage constant, and permittivity of the ceramic material. The influences of temperature, and

Figure 3.7 Center-mounted compression accelerometer

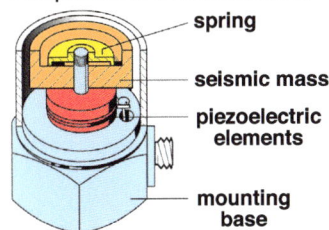

- spring
- seismic mass
- piezoelectric elements
- mounting base

Figure 3.8
Annular shear accelerometer

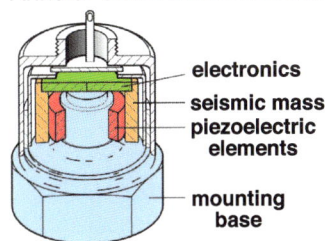

- electronics
- seismic mass
- piezoelectric elements
- mounting base

those of an additional source of error—bending of the mounting base—can be reduced by using a thick base.

Because center-mounted accelerometers do not eliminate all of the potential problems inherent to simple 33-mode accelerometers, other designs have been developed in the quest for highly reliable performance. An approach that has been used very successfully since the late 1960s, the annular shear design (Figure 3.8), is an elegant solution to some of the problems associated with 33-mode accelerometers. In the annular shear design, as in the center-mounted compression design, the piezoelectric element is an axially poled ring (tube) mounted around a central post. In an annular shear device, however, the electrodes are affixed to the cylindrical faces of the element, and the element operates in the shear (15) mode. The seismic mass also is a ring, mounted around the piezoelement.

This design eliminates pyroelectric charges. A shear piezoceramic element with electrodes perpendicular to the 1 direction will only produce an electric response to a strain /

stress in the form of shear in the 5 direction (i.e., $d_{11} = d_{12} = d_{13} = d_{14} = d_{16} = 0$, $d_{15} > 0$). In addition, the annular shear design reduces the effects of base bending and high sound pressure levels.

Since the introduction of the annular shear design, progressively more advanced designs have exploited the shear principle. Figure 3.9 shows three examples (all developed by Brüel & Kjær). In the DeltaShear® design* (Figure 3.9a), three piezo elements and three seismic masses are arranged in a triangular configuration around a triangular center post. In addition to allowing triaxial measurements, this design offers a high sensitivity-to-mass ratio, a relatively high resonance frequency, low weight, and effective isolation from base strains and transient temperatures.

In the patented ThetaShear® design* (Figure 3.9b), two piezoelectric elements are arranged in parallel between two posts and a seismic mass is positioned between the elements. A high-tension clamping ring holds all the components in place. This design is simple, is electrically insulated from the mounting surface, and has low mass loading.

The patented OrthoShear® design* (Figure 3.9c) incorporates a shear tube, like the annular ring design, but the outer electrode is divided vertically into four segments and is mounted to the base with hinges, allowing movement in all directions, so triaxial measurements can be performed. The single seismic mass is within the piezoceramic tube, making a very compact and reliable design. The x-, y-, and z-components of the acceleration are determined by summing the signals.

Figure 3.9 Shear-based accelerometers

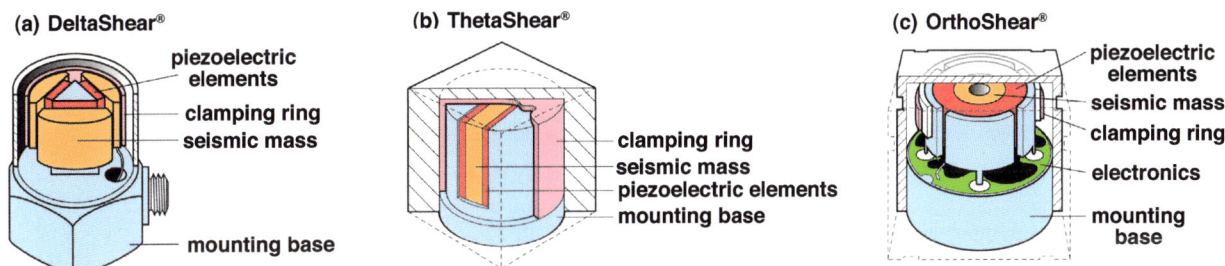

(a) DeltaShear®
- piezoelectric elements
- clamping ring
- seismic mass
- mounting base

(b) ThetaShear®
- clamping ring
- seismic mass
- piezoelectric elements
- mounting base

(c) OrthoShear®
- piezoelectric elements
- seismic mass
- clamping ring
- electronics
- mounting base

* DeltaShear,® ThetaShear,® and OrthoShear® are registered trademarks of Brüel & Kjær.
Figures 3.7, 3.8, 3.9 courtesy of Brüel & Kjær.

For accelerometer applications, soft piezo materials with relatively high piezoelectric voltage constants generally are preferred. Figure 3.10 plots g_{31} versus temperature for two soft PZT materials. Material II (a) is a Navy type II general-purpose material; material II (b) was developed especially for use in accelerometers. Note the weaker influence of temperature on the latter. Modified bismuth titanate materials are less sensitive than alternative materials but can be used at temperatures up to 550°C (Figure 3.11).

Flexional Sensors

To create the most commonly used type of flexional sensor, two plates or strips of piezoelectric ceramic, polarized in the third (thickness) direction, are bonded together to produce a flexible

Figure 3.10 Piezoelectric voltage constant versus temperature for two soft PZT materials

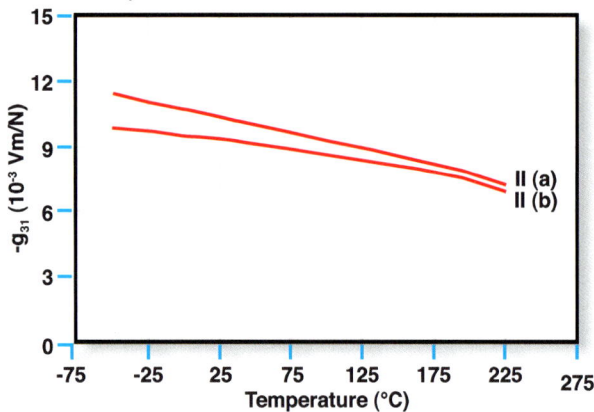

Figure 3.11 Permittivity versus temperature for modified bismuth titanate

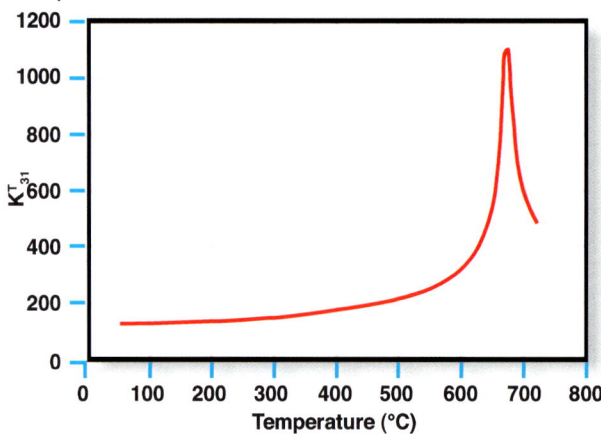

bilaminar element. In typical applications, a bilaminar element is secured in a cantilever mounting: one end of the element is fixed to a base and the physical force to be measured acts on the free end, as shown in Figure 3.12. If the two plates are configured with the direction of polarization opposing, and electrodes are affixed to their outer surfaces, as shown in Figure 3.12a, the element is a series element. If the two plates are configured with the same direction of polarization, the arrangement in Figure 3.12b produces a parallel element. For either a series bilaminar element or a parallel bilaminar element, one or the other of the two plates always will be subjected to a voltage opposite the polarizing voltage, so there is always a potential for electrically depolarizing the element. A bias voltage drive system, however, will apply the voltage to both plates of a parallel bilaminar element in the direction of polarization (Figure 3.12c). This will eliminate the possibility of electrical depolarization. Figures 3.13a and 3.13b show electrical circuits equivalent to a series bilaminar element and a parallel bilaminar element, respectively.

A perpendicular force at the free end of a bilaminar element bends the sensor, creating tension in the ceramic layer to which the force is applied and compression in the opposing layer. This physical input produces a voltage in each layer. In a series bilaminar element (polarity of plates opposing), the electric fields and voltages produced by the tensile and compressive forces have the same direction. The sum of the voltage generated by the layer under tension plus the voltage generated by the layer under compression indicate the extent of the physical force input. In a parallel bilaminar element (plates polarized in the same direction), the electric fields and voltages generated in the two plates are opposing, yielding larger displacement. Simpler, more economical construction makes the series bilaminar element the preferred configuration for a flexional sensor. The opposing polarization of the piezoelectric plates in series bilaminar elements also circumvents temperature-related problems, and this can be a significant advantage. On the other hand, the potential for electrically depolarizing the element

Figure 3.12
Flexional sensors (cantilever mounting)
(a) series bilaminar element

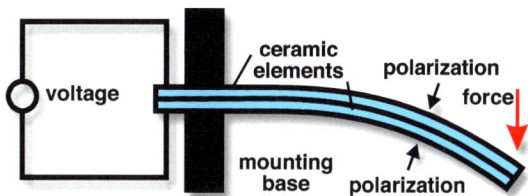

(b) parallel bilaminar element

(c) bias voltage drive ($V_2 \leq V_1$)

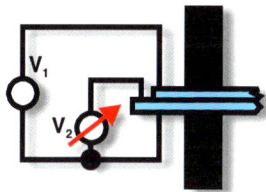

is always present.

A flexional piezoelectric element subjected to acceleration bends as shown in Figure 3.14. For a series bilaminar element, the output voltage, V, can be determined from:

Equation 3.15

$$V = 50 \, (l_f^3 / l_t) \, a$$

where

l_f = free length of bilaminar element

l_t = total length of bilaminar element

a = acceleration

A piezoelectric accelerometer can be made in miniature by coating a micromachined silicon cantilever with a thin film of lead titanate.

Relative to axial sensors, flexional sensors have much lower stiffness, lower mechanical impedance, and lower electrical impedance. Lower stiffness gives a flexional sensor a lower resonance (minimal impedance) frequency (Equation 3.9). Lower mechanical impedance and electrical

impedance make flexional sensors better adapted to sensing subtle mechanical input, and compatible with relatively simple amplifiers.

Stripe™ Actuators

Through composite layering technologies, APC International, Ltd. has developed the Stripe actuator, a flexional piezoelectric ceramic actuator that achieves greater deflections than conventional bilaminar elements in a wide range of sensor or actuator applications. In conventional cantilever mounting configurations, standard-sized Stripe actuators provide total deflections ranging from >0.7 mm to >2.5 mm, in response to a 150 V (maximum) input. The design of a Stripe actuator maximizes deflection, but, if movement is blocked, the actuator will develop a useable force, up to >0.40 N (Table 3.1). For detailed information about these flexional elements, see **Actuators**.

Time Constants

The capacitance of a flexional sensor element generally is high, and consequently the time constant is high. The time constant can be increased, if necessary, by connecting a capacitor in parallel with the flexional element. A parallel capacitor also will make the output voltage independent of temperature (see **Temperature Effects**), but will reduce sensitivity.

Figure 3.13 Electrical circuits equivalent to flexional sensors

(a) series bilaminar element   **(b) parallel bilaminar element**

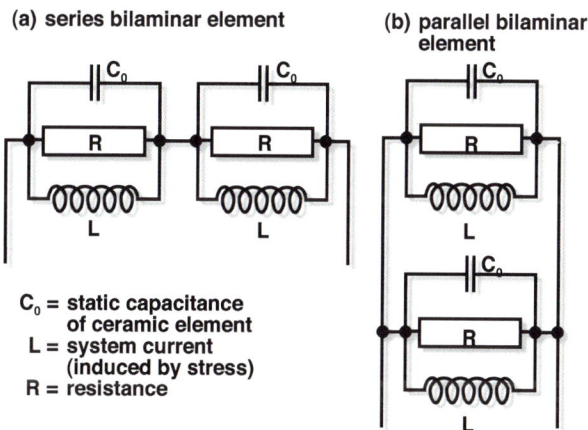

C_0 = static capacitance of ceramic element
L = system current (induced by stress)
R = resistance

Figure 3.14 Flexional acceleration sensor

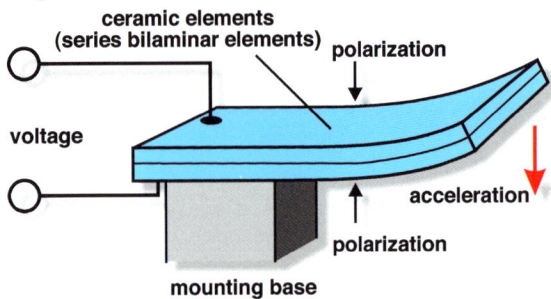

ceramic elements
(series bilaminar elements) polarization

voltage

acceleration

polarization

mounting base

Temperature Effects

Temperature Dependence of Piezoelectric Constants

The piezoelectric charge constant, d, the piezoelectric voltage constant, g, and the permittivity (dielectric constant), ε, are temperature dependent, but these relationships can be compensated for in the design of the sensor. Connecting a capacitor in parallel with the sensor eliminates the effects of temperature on output voltage. The increase in the total capacitance of the system will be accompanied by an equivalent reduction in the temperature coefficient for the total capacitance. Because $g = d / \varepsilon^T$ (ε^T = permittivity at constant stress), the temperature coefficient for g is assumed to be the difference between the temperature coefficients for d and ε^T, and output voltage will be essentially constant over a wide temperature range.

Pyroelectric Effects

Temperature fluctuations can cause a piezoelectric ceramic element to generate relatively high voltages, capable of disorienting the domains and depolarizing the element. In addition to obstructing measuring accuracy, such changes potentially can damage or destroy input stages of an amplifier. A capacitor incorporated into the circuit, in parallel with the sensor, will suppress the effects of temperature fluctuation by accepting the additional electrical energy. Series bilaminar elements circumvent temperature-related problems because the opposing polarization of the piezoelectric elements compensates for the pyroelectric effect.

Special-Purpose Designs and Applications

Composites

In sensor applications, the sensitivity of the ceramic element is maximal when the piezoelectric voltage constant, g, is maximal (1). For in-air applications, either the g_{33} constant (longitudinal mode constant) or the g_{15} constant (shear mode constant) of the element usually is put to use. In hydrostatic applications, however, sensitivity is proportional to constant g_h, which is equal to $g_{33} + 2g_{31}$ (1). For a piezoelectric ceramic element, $g_{33} \approx -2g_{31}$, thus the g_{33} and g_{31} contributions to g_h are subtractive, and the element exhibits poor sensitivity.

To counteract this problem and improve sensitivity, the sensor must be configured in a way that compensates for the effect of hydrostatic pressure. Traditionally, this has been accomplished by incorporating an air backing at one face of the ceramic element, incorporating air spaces into the sensor, or partially encapsulating the ceramic element in an absorbent polymer, to absorb some of the hydrostatic stress. A newer approach has been to use a composite structure—piezoelectric ceramic material admixed with a non-piezoelectric supporting material. The supporting material can be a polymer or, in some applications, a metal material, to maximize characteristics that contribute to sensitivity and minimize characteristics that detract from sensitivity.

By eliminating either the g_{31} or the g_{33} contribution to g_h, piezoelectric ceramic / polymer composites are particularly effective in underwater applications. A composite consisting of ceramic particles embedded in a matrix of rubber or polymer is called a 0-3 material, indicating that the ceramic particles are not in contact with one another, but that the supporting material is in contact with itself in three directions (see **Transducers**). The flexibility of the non-piezoelectric component allows the resulting material to be produced as highly flexible sheets or films, which are well suited for fabricating hydrophones (signal receivers). Alternatively, the ceramic particles can be embedded in a rigid polymer.

Table 3.1 Calculations for estimating performance characteristics of a Stripe actuator

Characteristic	Units	Calculation
total deflection	mm	$(2.2 \times 10^{-6})\,(l_f^2 / h^2)\,(V)$
compliance		$(26.4 \times 10^{-5})\,(l_f^3 / (w)\,(h^3))$
blocking force	N	deflection / compliance
resonance frequency	Hz	$(3.2 \times 10^5)\,(h / l_f^2)$

where l_f = free length of ceramic element
  h = height (thickness) of ceramic element
  V = voltage
  w = width of ceramic element

Observed values will be within ±30% of calculated values.

Composite materials in which the ceramic particles are in contact in one direction, such as fine fibers or thin rods of ceramic embedded in parallel in an inactive matrix, are called 1-3 materials. Like 0-3 materials, 1-3 materials can be fabricated in large and flexible sheets, according to the characteristics of the supporting matrix. Sensory capabilities of 0-3 and 1-3 materials are approximately equal, but 1-3 materials exhibit higher d_{33} values, a desirable characteristic for acoustic transducers.

Complex shapes, products with large surface area, and arrays can be constructed from 1-3 composites. Transducers for ultrasonic medical apparatus, nondestructive testing equipment for various materials, fish finders, and other devices are made by dividing a single ceramic plate into numerous small parts and embedding these parts in a polymer matrix. The shape of the transducer can be formed to a specific design tailored to the demands of the application.

0-3 piezoelectric ceramic / polymer composites exhibit high sensitivity, high pressure tolerance, a broad operating bandwidth, and a good acoustical impedance match for air or water. 1-3 piezoelectric ceramic / polymer composites have approximately the same characteristics as 0-3 composites, but are lighter and more rigid, and are more easily tailored to specific applications. Piezoelectric ceramic / metal composites (moonies and other flextensional devices) are designed such that their g_{33} and g_{31} constants additively contribute to g_h, rather than subtractively contribute, and these devices exhibit very high sensitivity. On the other hand, ceramic / metal composites are more pressure dependant and have much narrower operating bandwidths than ceramic / polymer composites.

Sensor Cables

A piezoelectric sensor cable consists of a copper wire core enveloped by unpolarized piezoelectric powder or polyvinylidene difluoride (PVDF) film and encased in an insulating sheath (Figure 3.15). Subjecting the cable to high voltage at a temperature near the Curie point for the piezoelectric material confers piezoelectric properties on the cable, enabling it to produce a signal when compressed. A principal application for sensor cables is for monitoring road traffic patterns—the number of signals recorded indicates the number of vehicles passing over the cable, differences in the extent of the responses can be used to categorize the vehicles by size.

Tactile Sensors

Typically constructed from thin sheets of piezoelectric materials, tactile sensors are used where force or pressure can be transferred between two surfaces in close proximity, such as in interactive "touch screen" displays or as a source of contact feedback in robotics applications (2). These highly sensitive sensors also have biomedical applications, in fields from dentistry to prosthetics. Some tactile sensors recognize static forces.

Figure 3.15 Sensor cable

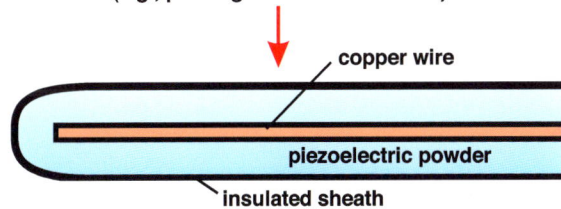

force (e.g., passing truck or automobile)
copper wire
piezoelectric powder
insulated sheath

In the tactile sensor shown in Figure 3.16, AC voltage from the oscillator causes the lower conducting layer (e.g., a PVDF film) to contract. The contractions in the lower layer are transmitted

Figure 3.16 Tactile sensor

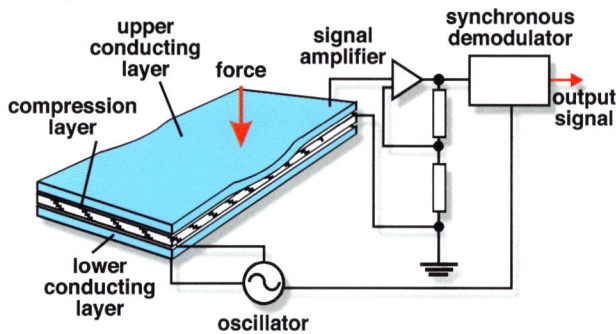

are in balance with the free charge carriers and dipoles in the element (Figure 3.18). The charge across the element is balanced and both elements generate the same electric charge. Voltage across the bias resistor is approximately zero. When gas movement increases, however, the charges at the surfaces of the element exposed to the gas are removed from the surface at an increased rate, and the charge across the element is thrown out of balance. A voltage across the resistor indicates the two elements are no longer in balance.

through the compression layer to the upper conducting layer, which responds to the mechanical stimulus by producing a voltage. When an external force is applied to the upper conducting layer, the demodulator recognizes a change in the signal pattern received from the three-layer assembly, and transmits a signal for processing.

Breeze Sensors

A breeze sensor discloses increases in the velocity of the air or other gas to which it is exposed. The sensor contains two piezoelectric elements in a series-opposed circuit (Figure 3.17). One element is exposed to the gas whose velocity is being monitored, the other is isolated from the gas. When the gas is motionless, or its flow is relatively constant, charged air molecules at the surfaces of the element exposed to the gas

Figure 3.18 Increase in gas flow unbalances charges in a breeze sensor

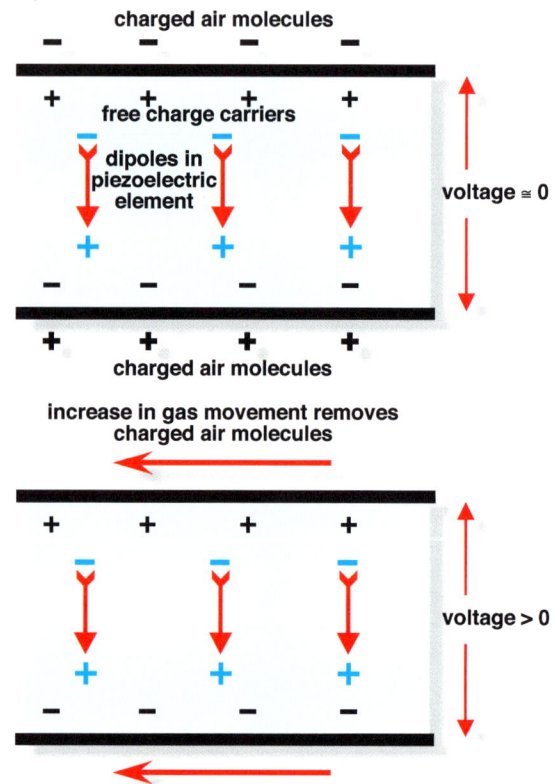

Figure 3.17 Breeze sensor incorporates two piezoelectric elements

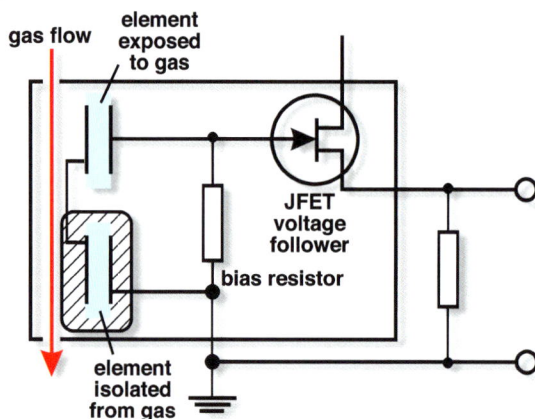

SAW Sensors

Surface acoustic waves (SAW), or Rayleigh waves (after the investigator who first predicted their existence) are a coupling of longitudinal and shear waves, passed along near the surface of a solid, but flexible, transmission medium. Surface acoustic waves have a number of applications in video signal transmission, and in sensory devices.

A SAW sensor is a sensitive gravimetric device that detects specific chemicals in a stream of gas. The action of a SAW sensor is analogous to that of a piezoelectric delay line (see **Transducers**). An electric oscillator causes a piezoelectric transmitter to flex, thereby generating mechanical waves in a solid, but flexible, transmission line, such as a silicon plate (Figure 3.19). The waves pass to a piezoelectric receiver, which converts them back to an electrical signal. The opposing surface of the transmission line, exposed to the gas stream, is coated with a layer of a target chemical-specific material. Interaction between the coating on the transmission line and the target chemical in the gas stream slows or accelerates the velocity of the mechanical waves as they pass from the transmitter to the receiver. The concentration of the target chemical in the gas can be quantified from the extent of the change in the signal.

By selecting the appropriate coating material for the transmission line, a SAW sensor can be designed to monitor a chosen chemical from among a wide range of gas-borne molecules. Applications to which a SAW sensor can be dedicated include monitoring target organic or inorganic molecules in air or monitoring the particulate composition of an aerosol or suspension. In the latter applications the SAW sensor can be used to ascertain particle sizes, as well as particle quantities. A well-designed and carefully constructed sensor can accurately detect nanogram quantities of the target chemical.

Figure 3.19 Surface acoustic wave sensor for detecting selected chemicals in a stream of gas

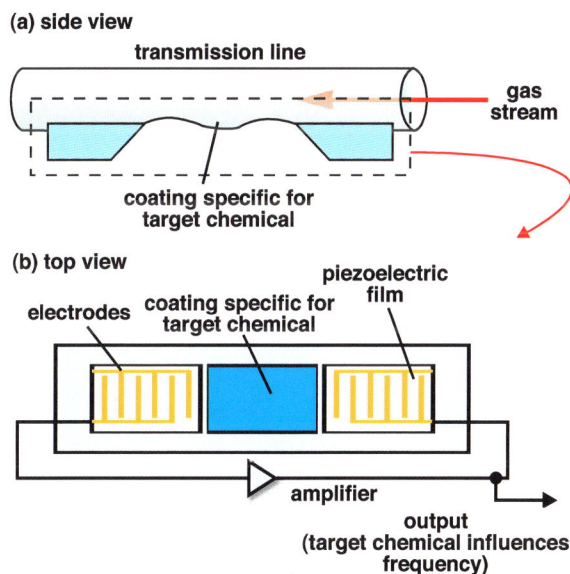

(a) side view

transmission line

gas stream

coating specific for target chemical

(b) top view

piezoelectric film

electrodes

coating specific for target chemical

amplifier

output (target chemical influences frequency)

References

1. Tressler, J.F. and K. Uchino, *Piezoelectric Composite Sensors* Vol. II/Appendix 37 in *Acoustic Transduction — Materials and Devices* Annual Report (1 January 1999 to 31 December 1999) Office of Naval Research, Contract No: N00014-96-1-1173 (June 2000). Request report from: Office of Naval Research, Regional Office Chicago, 536 S. Clark Street, Room 208, Chicago, IL 60605-1588.
2. Fraden, J. *Handbook of Modern Sensors* (2nd Ed.) American Institute of Physics Press, Woodbury, NY (1997).

actuators

A piezoelectric actuator converts an electrical signal into a precisely controlled physical displacement (*stroke*). If displacement is prevented, a useable force (*blocking force*) will develop. The precise movement control afforded by piezoelectric actuators is used to finely adjust machining tools, lenses, mirrors, or other equipment. Piezoelectric actuators also are used to control hydraulic valves, act as small-volume pumps or special-purpose motors, and in other applications requiring movement or force. Piezoelectric motors are unaffected by the energy efficiency losses that limit miniaturization of electro-magnetic motors—piezoelectric motors smaller than 1 cm^3 have been constructed. These motors also eliminate electromagnetic noise.

Although the stroke for a single piezoelectric ceramic element of typical dimensions (several millimeters thick) is measured in microns, the strokes for a number of such elements stacked together is additive.

The maximum strain for a stack actuator (the stroke divided by the height of the stack) can be 0.15% to 0.2% of the height of the stack.

A stack actuator is constructed in one of two ways: *discrete stacking* or *co-firing*. To prepare a discretely stacked actuator, the ceramic elements are individu-ally formed, fired, and polarized, electrodes are applied to each element, and the ceramic-electrode units are stacked together. Discrete stacking offers flexibility in designing the shape of the ceramic elements, and essentially all ceramic materials are amenable to the process. Moreover, discrete stacking ensures better control of the heat generated by the actuator during high frequency operation. On the other hand, the manufacturing process dictates that the individual ceramic layers be relatively thick, so the voltage needed to operate a discretely stacked actuator is high: 500-1000 volts.

To create an actuator that can be driven by lower voltages, each ceramic layer must be less than 1 mm thick. To achieve this thinness, the entire ceramic-electrode structure must be constructed and fired as a single unit (co-firing). Unfortunately, at present ceramics with high Curie points are rarely used to make co-fired actuators. As a consequence, most co-fired stacks cannot be operated at temperatures above 220°C, and most are incompatible with high electrical input.

A *flexional* or *bending actuator* is designed to produce a relatively large mechanical deflection in response to an electrical signal, similar to a flexional sensor. Two thin strips of piezoelectric ceramic are bonded together, usually with the direction of polarization coinciding, and are electrically connected in parallel. When electrical input is applied, one ceramic layer expands and the other contracts, causing the actuator to flex. Deflections are large, but blocking forces are low, relative to forces developed by stack actuators.

Flexional actuators are well suited to applications that demand sensitivity and a large response: to control needle movements in textile weaving machines, in switching applications, in Braille machines, and as small-volume pumps. A flexional actuator can open and close a valve, position a small lens or mirror, or maintain an electrical contact.

Actuators

When stimulated by an electrical signal, a piezoelectric actuator responds with a mechanical displacement (stroke) or, if displacement is prevented (blocked), with a force (blocking force) the magnitude of which is determined by the stiffness imparted by the design of the actuator. There is an approximately linear relationship between input signal and actuator response, either as a stroke or as a blocking force.

Figure 4.1 Piezoelectric actuators

(a) axial actuator

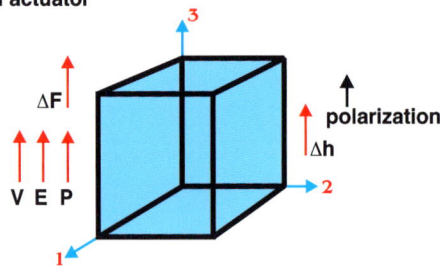

(b) transversal actuator

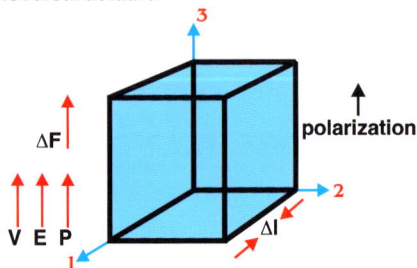

(c) flexional actuator (parallel bilaminar element)

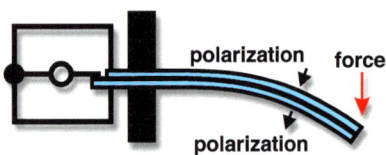

V = voltage
E = electrical field
P = polarization
h = thickness of ceramic element
l = length of ceramic element
F = applied force

Piezoelectric actuators are used to very precisely control positioning in applications ranging from low weight loads (e.g., optical equipment) to very heavy loads (e.g., machining tools), to dampen oscillations in tools, to generate high forces or pressures under static or higher frequency conditions, and as small but efficient motors. Short reaction times and fast acceleration rates make piezoelectric actuators attractive for fuel injection and valve control. In positioning applications, an infinitely small change in operating voltage produces an infinitely small change in position—positioning sensitivity is dictated by the ability to determine position, and by the limitations of the electronic supply device, not by the performance of the actuator. Piezoelectric actuators effectively control or negate vibration, even in challenging situations such as in wheeled vehicles and aircraft. Piezoelectric motors and pumps exhibit unique characteristics and in some applications (e.g., miniature motors) offer significant advantages relative to electromagnetic devices.

Like piezoelectric sensors, piezoelectric actuators are categorized according to the type of mechanical displacement they produce: axial, transversal, or flexional. An axial actuator (Figure 4.1a) accepts a signal applied parallel to the direction in which the piezoelectric ceramic element is polarized and creates a useable response—an extension—in the same direction (direction 3 input / direction 3 output, or d_{33}-mode). The ceramic element progressively extends in height as voltage is increased. A transversal actuator (Figure 4.1b) accepts a signal applied parallel to the direction in which the ceramic element is polarized and creates a useable response—a contraction—perpendicular to the direction of polarization (direction 3 input / direction 1 output, or d_{31}-mode). The ceramic element progressively shortens in length as voltage is increased. A flexional actuator (Figure 4.1c) is constructed from a bilaminar ceramic element (see **Sensors**). A flexional actuator operates in d_{31}-mode, like a transversal actuator, but its flexible construction makes it capable of significantly larger movement.

An axial, transversal, or flexional actuator also can be categorized by the complexity of its construction: a simple actuator consists of a single ceramic element; a compound actuator incorporates multiple ceramic elements stacked together and connected in parallel.

Axial and transversal actuators exhibit high stiffness and produce small movements or large blocking forces; flexional actuators are capable of larger movements but exert weak forces. Actuators are operated at various input frequencies, most commonly from static to approximately one half the resonance frequency of the mechanical system.

The basics of actuator design and behavior are described here, but no actuator is independent of its environment. Temperature, humidity, and other factors can affect the performance of an actuator, and must be carefully considered in choosing a model or developing a design. It is not unusual to find that an actuator ordered from stock, on the basis of characteristics measured under a narrow range of test conditions, does not perform as anticipated when challenged by a real application. Similarly, a power supply should not be chosen simply by matching its output capability to the drive requirements of the actuator. For best performance, the actuator and power supply should be considered interdependent, and should be chosen based on their combined suitability for a particular application.

Axial and Transversal Actuators

Simple Axial and Transversal Actuators

For a simple axial actuator consisting of a single ceramic element, under low electric field strength and low load, the stroke, blocking force, stiffness (spring constant), or resonance frequency can be determined from equations 4.1 – 4.6:

Stroke (force held constant):

Equation 4.1

$$\Delta h = d_{33} V$$

Blocking force (height of ceramic element held constant):

Equation 4.2

$$F_b = (d_{33} l w V) / (s^E_{33} h)$$

or, for a cylindrical element:

$$F_b = (d_{33} \pi r^2 V) / (s^E_{33} h)$$

Stiffness:

Equation 4.3

$$K_E = \Delta F / \Delta h$$

or

Equation 4.4

$$K_E = (lw) / (s^E_{33} h)$$

or, for a cylindrical element:

$$K_E = (\pi r^2) / (s^E_{33} h)$$

Minimum impedance (resonance) frequency (free element, l and w both $\leq$ h):

Equation 4.5

$$f_m = N_L / h$$

Minimum impedance frequency (one surface of element secured to a base):

Equation 4.6

$$f_m = N_L / 2h$$

For a transversal actuator, determine corresponding values from equations 4.7 – 4.12:

Stroke (force held constant):

Equation 4.7

$$\Delta l = (d_{31}) \, (l \, / \, h) \, (V)$$

Blocking force (length of ceramic element held constant):

Equation 4.8

$$F_b = (d_{31} / s^E_{11}) \, (w) \, (V)$$

Stiffness:

Equation 4.9

$$K_E = \Delta F / \Delta l$$

or

Equation 4.10

$$K_E = (hw) / (s^E_{11} l)$$

Minimum impedance (resonance) frequency (free element):

Equation 4.11

$$f_m = N_P / l$$

Minimum impedance frequency (one surface of element secured to a base):

Equation 4.12

$$f_m = N_P / 2l$$

where

F = force

h = height (thickness) of ceramic element

d_{33}, d_{31} = piezoelectric charge constants*

V = voltage

F_b = blocking force

w = width of ceramic element

s^E_{33}, s^E_{11} = elastic compliance (constant electric field)**

K_E = stiffness of ceramic material

l = length of ceramic element

f_m = minimum impedance frequency (resonance frequency) of ceramic element

N_L = longitudinal frequency constant for ceramic material

N_P = planar (radial) frequency constant for ceramic material

* electric polarization generated per unit of mechanical stress applied

** strain produced per unit of mechanical stress applied

Piezoelectric charge constants d_{33} and d_{31} typically increase as the electric field strength is increased and, consequently, at high field strength the stroke for an actuator can exceed the value anticipated from equation 4.1 or 4.7. Similarly, the blocking force can exceed the value anticipated from equation 4.2 or 4.8, but the increase will not be proportionally as great as the increase for the stroke. Further, because the charge constants also are temperature dependent, the stroke of an actuator can increase with increasing temperature. Of course, the extent of these changes also depends on the characteristics of the ceramic used to construct the actuator. The manner in which the actuator is mounted (and pre-stressed, if applicable) can further complicate the situation. Consequently, measurements made under anticipated operating conditions are always the most reliable indicators of actuator performance.

Figure 4.2 Influence of constant and spring forces on an axial actuator

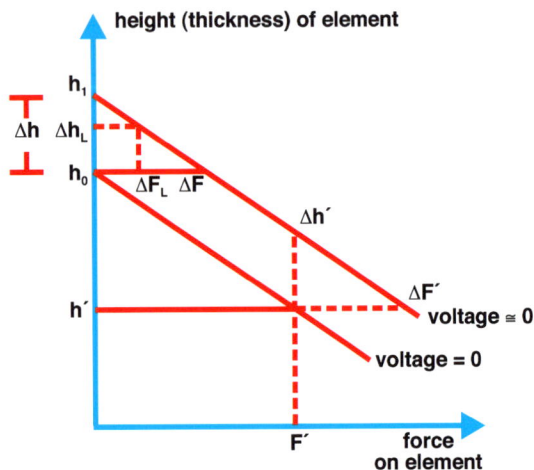

Exertion of a constant force on a single element axial actuator will decrease the length of the piezoelectric element to an extent dictated by the magnitude of the force and the stiffness of the element. If hysteresis effects are absent, or negated, the relationship between the force on the actuator and the resulting stroke or blocking force will be linear, up to very high loads. Figure 4.2 shows these relationships. In the absence of an applied force or voltage, the height (thickness) of the ceramic element, h, is h_0. Applying a voltage, V, increases h to $h_0 + \Delta h = h_1$. A constant force, F_C, shifts the baseline. Under these conditions voltage V causes the element to increase in height by $\Delta h'$, which almost equals Δh, the increase in height in the absence of the exerted force.

In some actuator applications, however, the force exerted on the ceramic element is applied by a spring, and the force increases as the ceramic element expands. Variable force on the element makes a significant difference in the force:stroke or force:blocking force relationship, depending on the spring constant:

Equation 4.13

$$\Delta h_L = \Delta h / (1 + (K_L / K_E))$$

Equation 4.14

$$\Delta F_L = \Delta F_C / (1 + (K_L / K_E))$$

where

h_L = stroke associated with spring force
h = height (thickness) of ceramic element
K_E = stiffness of ceramic material
K_L = spring constant
F_L = spring force
F_C = constant force

In conjunction with a spring force, F_L, voltage V produces stroke Δh_L, which is smaller than $\Delta h'$, the stroke voltage V would produce in the presence of a constant force.

Compound (Stack) Axial Actuators

Any type of actuator enhances certain desired features at the expense of others (e.g., the design of a flexional actuator maximizes movement but

Figure 4.3
Compound axial actuator (stack actuator)

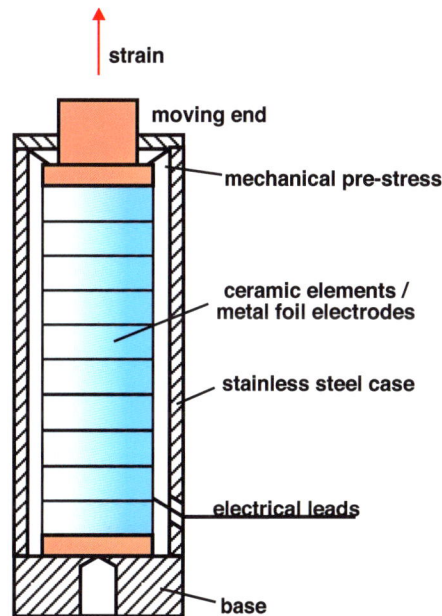

strain
moving end
mechanical pre-stress
ceramic elements / metal foil electrodes
stainless steel case
electrical leads
base

sacrifices force). Selection of a suitable actuator depends on knowing which desirable features are most important to the application at hand or, conversely, knowing which drawbacks must be minimized. For most applications, simple axial actuators have little practical value, because the stroke for a single element is too small, or because a very high voltage—typically kilovolts—is needed to produce a useable stroke. Consequently, far more widely used are compound, or stack, axial actuators. A stack actuator is constructed of multiple ceramic elements stacked together, with the direction of polarization of each element parallel to the direction in which the stroke or blocking force will be developed, and connected in parallel (Figure 4.3).

A stack actuator can be designed for operating at 100 v or less, and its maximum stroke can approach 0.2% of the height of the stack of elements. Stroke and other properties of multiple element axial actuators can be determined from equations 4.1 through 4.6, by substituting nV for V or nh for h, with n representing the number of ceramic elements in the stack.

A stack actuator is constructed in one of two ways: discrete stacking or co-firing. To prepare a discretely stacked actuator, each ceramic element is individually formed, fired, and polarized, electrodes are affixed to each element, and the ceramic-electrode units are stacked together, alternating with layers of thin metal sheeting through which electrical contact is made. Discrete stacking offers flexibility in the design of the shape of the ceramic elements, and the widest range of ceramic materials is amenable to the process. Moreover, discrete stacking ensures better control of the heat generated by the actuator during high frequency operation. On the other hand, the thinness to which individual ceramic layers can be manufactured, fired, and assembled is limited. As a consequence, the voltage needed to operate a discretely stacked actuator, although less than the kilovolts of input needed to maximize the stroke for a single element actuator, still is high. Because a 500-1000 V input is needed to operate a discretely stacked actuator, actuators produced by this process often are called high-voltage actuators or high-power actuators.

To create an actuator that can be driven by lower voltages, each ceramic layer must be considerably less than 1 mm thick. To achieve this level of layer thinness, the entire ceramic-electrode structure must be constructed and fired as a single unit (co-firing). In theory, any piezoelec-

Table 4.1 Stroke increases in proportion to an increase in stack height; blocking force increases in proportion to an increase in stack cross-sectional area

Stack Height	Stack Cross-Section	Stroke	Blocking Force
h	A	1	1
2h	A	2	1
h	2A	1	2
2h	2A	2	2

tric ceramic can be used to prepare a co-fired stack actuator, but at present ceramics with high Curie points are rarely used for this purpose. As a consequence, most commercially prepared co-fired stacks cannot be operated at temperatures above 220°C, nor are they compatible with high electrical input.

The stroke for a stack actuator is proportional to the height of the stack of elements (Table 4.1). Depending on the ceramic material used to construct the stack, the maximum strain (stroke / stack height) can equal up to 0.15% to 0.2% of the height of the stack. The blocking force developed by a stack is proportional to the cross-sectional area of the stack (Table 4.1). There also is an important relationship between strain and power consumption: the power requirement of an actuator increases by a factor of two in proportion with increases in strain. Consequently, operating an actuator of stack height (h) at half its potential strain, compared to operating a stack of height (h / 2) at full strain, will reduce power consumption by 50%. The taller actuator exerts half the strain / voltage, but there is a cost—capacitance is twice as high. When a stack actuator must generate a stroke against a varying force—a spring, a clamping mechanism, or a varying mass load—or when it is used to generate vibrations or pulses, the reaction by the actuator will vary, and will depend on the relationship between the applied force and the stiffness of the actuator.

The useful properties of a stack actuator are determined by the characteristics of the ceramic material, the manner in which the stack is constructed (discretely stacked or co-fired), and

Stripe actuators

the height and cross-sectional area of the stack. The stack of ceramic elements can be simply coated with an insulating material (bare stack) or, for protection from mechanical damage, humidity, or other environmental effects, or to incorporate mechanical pre-stress, the stack can be encased in a stainless steel or other casing. Pre-stress is a necessary component of an actuator intended for dynamic operation, for example, to compensate for the high tensile forces developed under these conditions.

Construction of a stacked axial actuator involves additional compromises. The stacked element construction can significantly reduce the rigidity of the system, and the desirable rigidity of a single element actuator can be approached only by incorporating a mechanical pre-stress, typically via a spring or a central bolt. Also, because the ceramic elements have low tensile strength, a compound actuator that does not incorporate pre-stress must be protected from tensile forces in some other way, especially if it is to be used in pulse operations.

If a highly elastic spring is used to supply the pre-stress, the stroke, blocking force, and stiffness will be as predicted in equations 4.1 – 4.3, but the additional mass will prolong the response time. On the other hand, a rigid spring or central bolt will make the stack more rigid, as desired, but will act as a loading spring and will reduce the performance of the actuator. High mechanical pre-stress also can lead to partial depolarization of the piezoelectric elements. Bonding the elements will eliminate the drawbacks associated with pre-stressing the stack, but the layers of bonding material will make the actuator more elastic, and vulnerable to tensile forces.

Strain cannot be the only consideration when constructing or selecting a stack actuator. Selection of a suitable ceramic material for a particular application involves determining the best balance of electromechanical characteristics, including power consumption, the amount of heat generated during operation, and the intended operating temperature. In an actuator intended for high frequency operation, for example, hysteresis losses in the ceramic material could heat the actuator unacceptably, and must be compensated for (e.g., through a feedback-controlled system). Also, extraneous resonances below the resonance frequency of the actuator could be introduced by the pre-stress mechanism, the housing, or other sources. To avoid this complication, the system should be made as compact as possible (e.g., by using a ceramic with high strain characteristics). For a particular ceramic, strain, elasticity, dynamic response, resonance frequencies, and other properties will be consistent, whether the ceramic is used in co-fired or discretely stacked actuators.

Multilayer Actuators

An axial or transversal actuator can be constructed from multiple layers of millimeter-thick to sub-millimeter-thick ceramic (Figure 4.4). Displacement per unit height or length of ceramic in a multilayer actuator is equivalent to that for a conventional stacked element actuator, so there is no displacement advantage to a multilayer actuator, but movement is achieved at much lower voltage: a 50 V input can provide an electric field of approximately 1 kV/mm. Despite this advantage, practical voltages induce an increase in thickness of only a few micrometers for each layer of a multilayer actuator. Thus, to obtain a useable stroke it usually is necessary to stack several multilayer axial actuators.

Figure 4.4 Multilayer actuators

(a) axial actuator

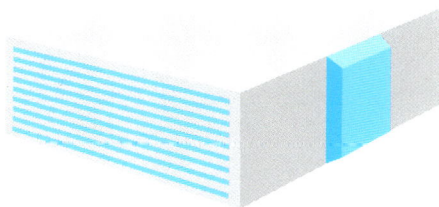

(b) transversal actuator

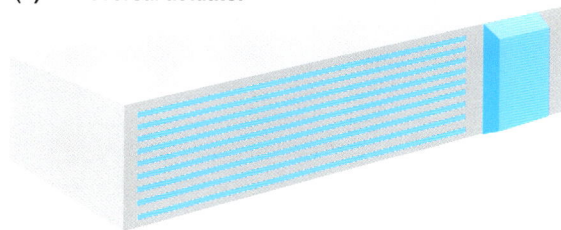

Multilayer Pseudoshear Actuator

Investigators at the Pennsylvania State University [1] stacked polarized rectangular sheets of piezoelectric ceramic and, using a stiff, conductive epoxy, bonded the sheets at alternate ends and bonded one end of the bottom layer to a fixed base (Figure 4.5). For stability, the gaps between the sheets were filled with thin plastic sheets. The ceramic sheets were electrically connected in parallel through the conductive bonding material.

When an electric field counter to the direction of poling is applied to the bottom sheet, alternate sheets are exposed to a negative or a positive electric field, and elongate or shrink, respectively, but all displacement is in the same direction. As a result, the actuator displays a strong shear motion about the face perpendicular to the direction of bonding (Figure 4.5). The net result simulates the performance of a shear actuator, hence the name of the device. If the voltage at the bottom sheet is reversed, the direction of the displacement is reversed (Figure 4.5).

The elongation, Δl_e, or shrinkage, Δl_s, of each alternating layer is given by:

Equation 4.15

$$\Delta l_e = d_{31}E_3 l$$

or

$$\Delta l_s = -d_{31}E_3 l$$

where

d_{31} = piezoelectric charge constant
E = electric field
l = length of the ceramic sheet

Total displacement, Δl_{tot}, at the bonded end or the free end of the top sheet, respectively, is:

Equation 4.16

$$\Delta l_{tot} = (n-1)d_{31}E_3 l$$

or

$$\Delta l_{tot} = (n)d_{31}E_3 l$$

where

n = number of sheets in the stack

Figure 4.5 Multilayer pseudoshear actuator

(a) at rest

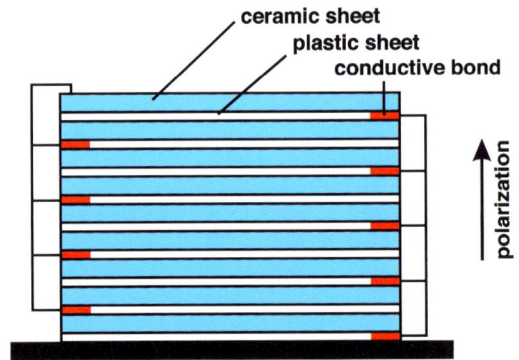

(b) negative field at bottom sheet

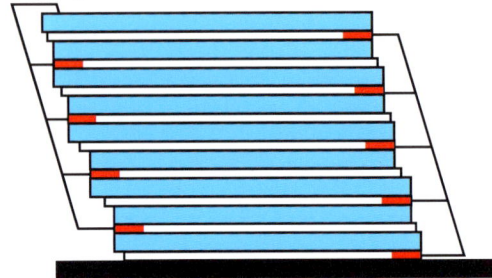

(c) positive field at bottom sheet

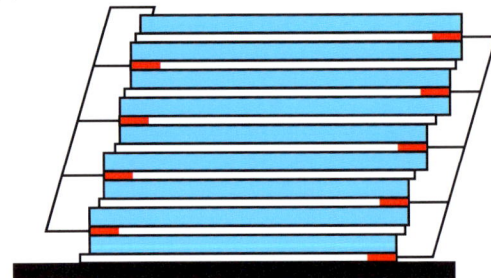

In an alternating current, the actuator vibrates in a manner similar to true shear vibration.

The investigators achieved more than a 50 μm displacement from an actuator composed of eighteen 25.57 x 4.02 x 0.51 mm (l x w x h) layers. Displacement can be increased and driving voltage can be reduced, while maintaining a constant electric field strength, by reducing the thickness of the sheets and increasing the number of sheets. Devices of this kind could replace conventional axial actuators in many applications, including motor applications, flow sensing and flow control, valving or pumping systems, and other applications.

Mounting a Stack Actuator

Proper coupling of a stack actuator to a mechanical system will transfer only axial forces, and will preclude any bending, tilting, or shear forces that could inhibit the stroke or damage the actuator (Figure 4.6). A bare (unencased) stack actuator must be mounted only at its endfaces—clamping the stack at its circumference could damage the ceramic components or the electrodes. An encased stack can be mounted at the endfaces or clamped at the circumference. Typically, the end of the actuator closest to the electrical lead is used as the mounting surface.

Suitable coupling often is achieved by creating an interface that combines a ball tip end and a planar face end (Figure 4.6a), with the point of contact at the center of the planar face. Alternatively, planar face to planar face contact is acceptable, if one of the faces is allowed free movement (Figure 4.6b). Without this freedom, any misalignment between the faces will create mechanical overstress at the edge of the stack (Figure 4.6c). In addition to protection from mechanical damage or environmental effects, another advantage to an encased and pre-stressed actuator, relative to a bare stack actuator, is that shear forces will be directed to the pre-stress mechanism, and the ceramic elements will be not be subjected to these forces.

If the stroke of a particular stack actuator is insufficient for a specific application, a longer actuator, or multiple actuators in series, might be considered. It might not be necessary to resort to these measures, however—a mounting configuration such as one of those shown in Figure 4.6d often can effectively amplify the stroke. Alternatively, because the power required by an actuator increases in proportion with increasing strain, there can be advantages to using an overly long high-voltage actuator in combination with a lower-voltage power supply. The power supply will drive the actuator to only a fraction of its potential strain, but this combination will significantly reduce power consumption relative to using a high-voltage power supply to drive a shorter high-voltage actuator to full strain. Also important—sometimes more important—is the fact that the combination of long high-voltage actuator and low-voltage power supply will minimize self-heating if the actuator is operated under dynamic conditions. These advantages often overcome the disadvantages entailed by using a longer actuator—larger capacitance, lower stiffness, and a lower resonance frequency. Advantages and drawbacks of this approach should be evaluated according to the demands of the application in question. As stated at the beginning of this chapter, the actuator and power supply should be chosen based on their combined suitability for a particular application.

When a stack actuator is to be used to ensure highly precise and stable positioning, the actuator should be used in combination with a sensor, in a closed loop (feedback) system. A strain gauge or an optical sensor, inductive sensor, or interferometer can be used to monitor the

Figure 4.6 Interface between stack actuator and mechanical system

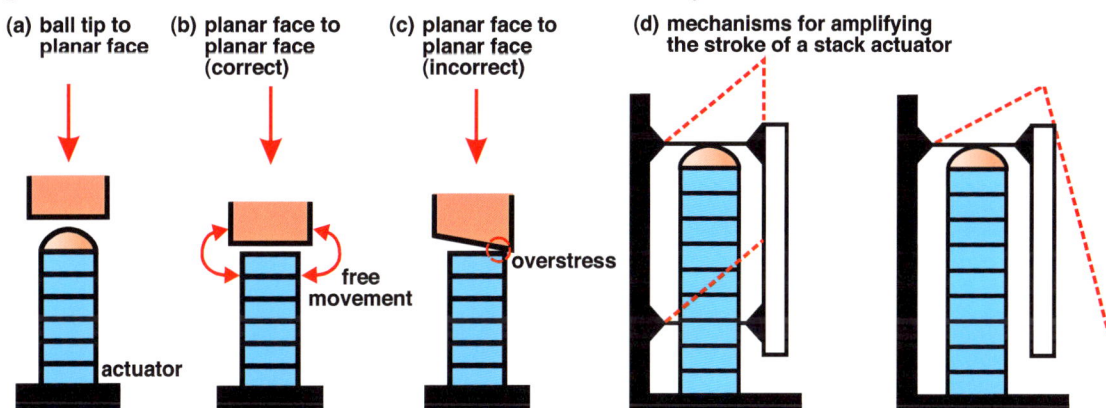

(a) ball tip to planar face (b) planar face to planar face (correct) (c) planar face to planar face (incorrect) (d) mechanisms for amplifying the stroke of a stack actuator

free movement

overstress

actuator

absolute position of the actuator and, if necessary, adjust the voltage. A stack actuator will convert an infinitely small change in input voltage into an infinitely small mechanical displacement, down to displacements at atomic levels, but precise determination of position is critical to precise feedback-based control, and extremely fine control of position cannot be obtained by applying a fixed voltage to the actuator. The varying load acting on the actuator will compress the actuator at a very small, but variable, level, and will shift its position in a manner that cannot be anticipated nor compensated for by applying a fixed voltage. It is critical to determine the exact position of the actuator and finely adjust the input voltage to correct or maintain that position.

Because the actuator is capable of infinitely small mechanical displacement, position control is governed by the limitations of the position-sensing and power-supplying apparatus. A strain gauge, applied to the surface of an actuator, is an effective and inexpensive means of determining the actual position (displacement) of the actuator. The strain gauge will detect the product of the various factors affecting the actuator's displacement, including the input voltage, the (varying) load, and creep and hysteresis effects in the piezoelectric elements. Position detection will

Figure 4.8 Charge drive circuit for piezoelectric actuator

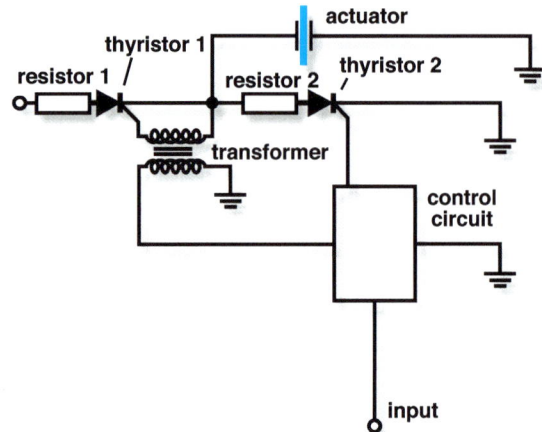

be accurate to within approximately 0.1% of the maximum displacement of the actuator, at microsecond-level response times.

A sensitive position-sensing Wheatstone bridge configuration is shown in Figure 4.7. Usually, two of the strain gauges (e.g., resistor 2 and resistor 4) will be affixed parallel to the axis of the piezoelectric stack and two will be affixed perpendicular to the axis. When (impedance 1 / impedance 2) = (impedance 3 / impedance 4) the bridge is in balance (2). A change in one or more impedance unbalances the bridge and creates a positive or negative output voltage, which can be corrected for via a feedback control circuit. Linking an actuator to a sensor of this type, calibrated against an absolute measuring system, assures linear and reproducible displacements by the actuator. Some manufacturers incorporate sensors into the configurations of their actuators, thereby eliminating the need to add components (and volume) to the system.

Note that additional factors affect positioning if a mechanical arrangement such as one of those shown in Figure 4.6d is used as an intermediary between the displacement of the actuator and the action point. In such a system the strain gauge-based sensor should be replaced by an eddy current sensor, a capacitive sensor, or an inductive sensor (linear variable differential transformer) (2), and conditions should be monitored at the action point, rather than at the surface of the actuator.

Figure 4.7 Four-gauge Wheatstone bridge for sensing position (strain) of a piezoelectric actuator

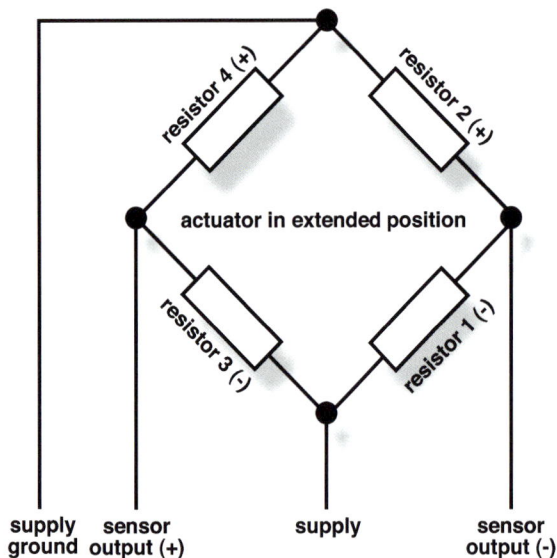

Figure 4.9
Parallel bilaminar flexional actuator

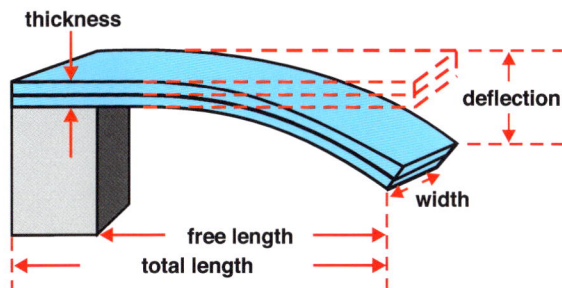

Circuitry

The voltage and power output from a signal generator usually cannot be directly input into a piezoelectric actuator, and an amplifier must be used to convey the displacement requirements (extent and frequency) to the actuator. As discussed previously in **Mounting a Stack Actuator**, the properties of the driving electronics are highly important to determining the suitability of an actuator for a specific purpose. In fact, an actuator can be used in widely disparate applications simply by changing the nature of the power source. Increasingly stringent demands on the actuator dictate increasingly larger power input.

The low voltages characteristic of operating multiple element actuators are counterbalanced by high capacitances and high operating currents. To be compatible with the fast-response capabilities of a high power actuator, the electrical components in the system must be capable of delivering high currents, at appropriate voltages, during short periods. The two alternatives for driving an actuator are voltage drive circuitry and charge drive circuitry. Voltage drive circuitry is simpler, but a feedback controlled system will be required to compensate for hysteresis and creep (time-associated deterioration in displacement for a specific applied voltage) inherent to this option. In contrast, charge drive circuitry will provide an almost linear relationship between stroke and dielectric displacement.

Figure 4.8 shows a basic charge drive circuit. The actuator is charged via thyristor 1 and current-limiting resistor 1, and is discharged via thyristor 2 and resistor 2. The circuit is compatible with high currents, so switching can be extremely rapid.

In either a charge drive system or a voltage drive system analog drive offers a valuable advantage: it allows continuous displacement of the actuator, between zero displacement and full displacement, rather than simply driving the actuator from the fully contracted position to full displacement. This control is achieved by comparing the voltage to the actuator against a preset voltage in the regulator stage of an amplifier, and charging or discharging the actuator by the variable amounts required to maintain the voltages in balance.

Figure 4.10   Electrical connections for bilaminar flexional actuators

(a) parallel drive

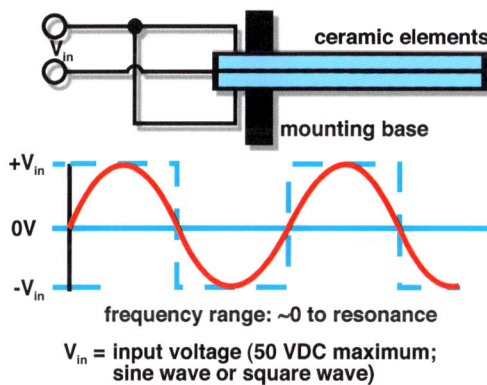

(b) bias drive

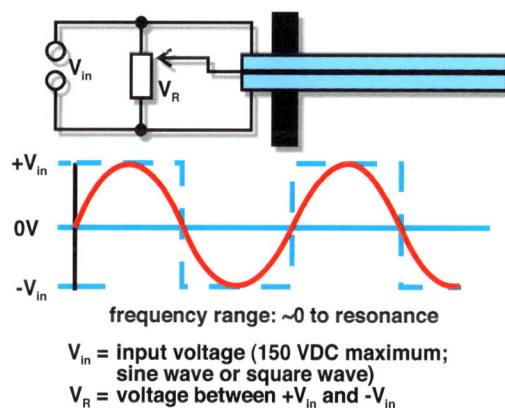

Flexional Actuators

Bilaminar piezoelectric ceramic elements are widely used as sensors, but they also are responsive actuators. Typically, two thin ceramic strips or plates are bonded together, with electrodes affixed to each layer to create a parallel bilaminar element (layers polarized in the same direction, electrically connected in parallel). When voltage is applied, one layer expands and the other actively or passively contracts, causing the actuator to bend (Figure 4.9). Series bilaminar elements offer important advantages for sensor applications (see **Sensors**), but for actuator applications parallel bilaminar elements (Figure 4.10a) offer higher sensitivity. Bias voltage circuitry (Figure 4.10b) will generate an electric field parallel to the direction of polarization, eliminating the potential for depolarizing the ceramic.

Deflections with a flexional actuator can be high, but blocking forces will be low, relative to forces achieved by axial or transversal actuators. The ceramic elements also usually exhibit relatively low resonance frequencies. Resonance frequency can be increased by making the actuator element thinner.

Stripe™ Actuators

Through composite layering technologies, APC International, Ltd. has developed the Stripe actuator (Figure 4.11), a flexional piezoelectric ceramic actuator that achieves greater deflections than conventional bilaminar actuators. In conventional cantilever mounting configurations, the four standard sizes of Stripe actuators provide total deflections ranging from > 1.1 mm to 2.6

Figure 4.11 Stripe actuator: a better bending actuator

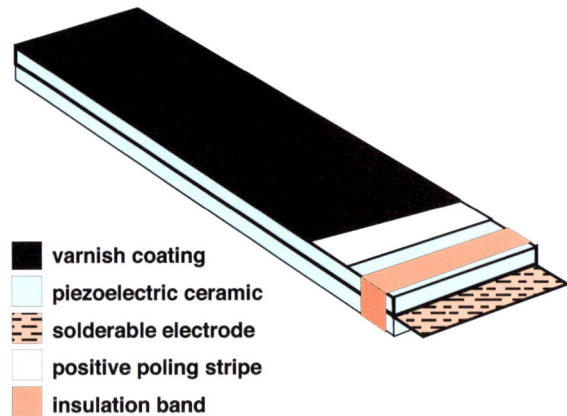

- ■ varnish coating
- ▨ piezoelectric ceramic
- ▤ solderable electrode
- □ positive poling stripe
- ▥ insulation band

- Proprietary layering technology increases flexibility, allows greater deflection
- Parallel electrical configuration of ceramic layers ensures high sensitivity to input; compatible with bias voltage circuitry that eliminates the potential for depolarizing the ceramic layers
- Varnish layer electrically insulates surface, protects from humidity, dust, other hazards
- White stripe identifies positive surface
- Solderable electrode bonded between plates

mm, in response to the maximum input voltage. (Table 4.2). The design of a Stripe actuator maximizes deflection, but, if movement is blocked, the actuator will develop a useable force, up to > 0.60 N (Table 4.2, Figure 4.12). The relationships in Table 4.3 can be used to estimate the deflection and blocking force for a Stripe actuator that is mounted with a free length other than listed in Table 4.2. Observed values will be within ±30% of calculated values.

Table 4.2 Characteristics of Stripe actuators

Stripe Actuator	Dimensions (mm)			Free Length (mm)	Total Deflection (mm)	Blocking Force (N)	Resonance Frequency (Hz)	Capacitance (pF)
	Length	Width	Thickness					
600/200/0.70SA	60.0	20.0	0.70	50	> 2.6	> 0.50	~60	190,000
490/021/0.80SA	49.0	2.0	0.80	38	> 1.6	> 0.11	~110	10,000
400/200/0.70SA	40.0	20.0	0.70	30	> 1.1	> 0.60	~170	120,000
350/025/0.70SA	35.0	2.5	0.70	25	> 1.1	> 0.16	~250	15,000

Deflection and blocking forces measured under driving voltage of 150 V in direction of polarization.
Operating temperature: -25°C to 70°C; storage temperature: -40°C to 85°C.

Figure 4.12 Blocking forces and (half) deflections for Stripe actuators

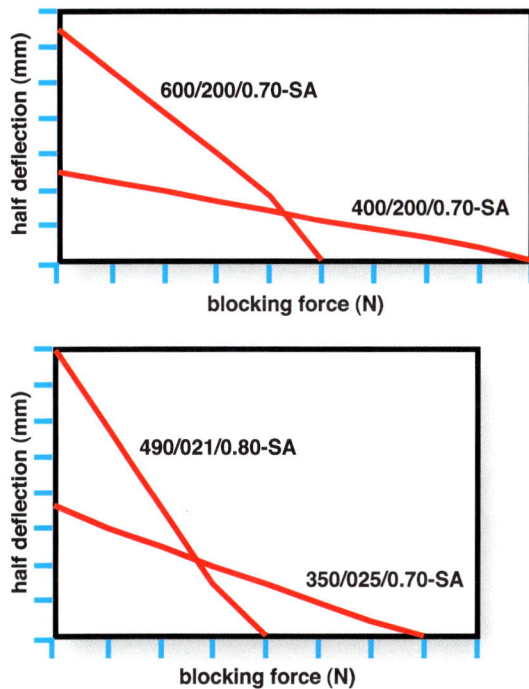

half deflection (mm)

600/200/0.70-SA

400/200/0.70-SA

blocking force (N)

half deflection (mm)

490/021/0.80-SA

350/025/0.70-SA

blocking force (N)

Table 4.3 Calculations for estimating performance characteristics of a Stripe actuator

Characteristic	Units	Calculation
total deflection	mm	$(2.2 \times 10^{-6})\,(l_f^2 / h^2)\,(V)$
compliance		$(26.4 \times 10^{-5})\,(l_f^3 / (w)\,(h^3))$
blocking force	N	deflection / compliance
resonance frequency	Hz	$(3.2 \times 10^5)\,(h / l_f^2)$

where l_f = free length of ceramic element
h = height (thickness) of ceramic element
V = voltage
w = width of ceramic element

Observed values will be within ±30% of calculated values.

Stripe actuators derive their name from a white stripe that identifies the positive surface. Both plates of a Stripe actuator are polarized in the same direction (parallel bilaminar element), and a solderable electrode is bonded between the plates. Vulnerable surfaces of a Stripe actuator are coated with a special varnish that electrically insulates the surface and protects it from humidity, dust, and other hazards that can shorten the working lifetime of an unprotected actuator.

Widthwise Bending Actuator
Unlike conventional bilaminar bending actuators, in which the deflection is in the thickness direction, a recently developed multilayer actuator was designed to bend in the widthwise direction (3). Compared to a bilaminar bending actuator of similar dimensions, this multilayer actuator exhibits a significantly higher resonance frequency, and produces a displacement twice as large and a blocking force more than 27 times as large. These characteristics make a widthwise flexional actuator very appealing for use in the servo control system of the magnetic head tracking mechanism in computer hard disk drives.

Multilayer Flexional Actuators
Multilayer flexional actuators (Figure 4.13) are constructed by combining electrodes and layers of polarized ceramic in a manner that enables the layers to expand or contract like a single bilaminar element. Like multilayer axial and transversal actuators, multilayer flexional actuators typically require much lower drive voltages, compared to their bilaminar equivalents. Additional advantages of multilayer flexional actuators, relative to simple bilaminar actuators, are summarized in Table 4.4. The maximum strain for multilayer devices is the same for multilayer and single element flexional actuators; general rules and equations governing conventional flexional actuators apply to multilayer devices as well.

Figure 4.13 Multilayer flexional actuator
(parallel construction)

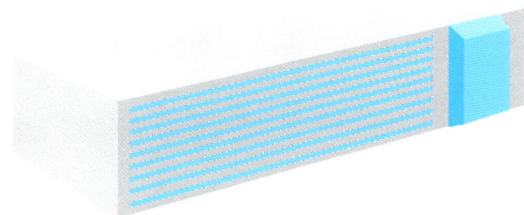

Table 4.4
Characteristics of multilayer flexional actuators and bilaminar actuators compared

Characteristic	Multilayer Actuator*	Bilaminar Actuator
Displacement	small (10 μm)	large (300 μm)
Force	strong (1000 N)	weak (1 N)
Response Speed	fast (10 μs)	slow (1 ms)
Lifetime	longer (10^{11} cycles)	shorter (10^8 cycles)
Electromechanical Coupling Factor	large (k_{33} = 70%)	small (k_{33} = 10%)

Approximately 100 ceramic layers.
Data from (5, 6).

Flextensional Devices

Flextensional devices combine desirable characteristics of multilayer actuators with those of bilaminar actuators. These devices, nicknamed "moonies" or "cymbals", according to their design (Figure 4.14) transform a small thickness mode displacement by a multilayer ceramic element into flexional movement by two metal plates that encompass the element. Through the mechanical gain from this transfer, a flextensional device will exhibit multiple times greater displacement than could be achieved by the multilayer actuator element alone, and significantly greater force and faster response than an equivalent bilaminar actuator. The "cymbal" modification can significantly increase the displacement, relative to the basic design. Designs other than those shown in Figure 4.14 have been developed. Flextensional devices typically are designed with resonance frequencies of 300 to 3000 Hz.

Applications for Piezoelectric Actuators

If cost were the only measure of practicality, piezoelectric actuators currently would be at a disadvantage, compared to electromechanical alternatives. In a number of applications, however, the characteristics of piezoelectric actuators provide a performance advantage—or advantages—that cannot be matched by electromechanical devices. Several applications and potential applications for piezoelectric actuators are illustrated in Figure 4.15. Motors and pumps are described here.

Piezoelectric Motors

There is an increasing need for small to extremely small (<1 cm³) motors in industrial, commercial, medical, and other applications, but electromagnetic motors of this size are not energy efficient. Ultrasonic piezoelectric motors are an ideal alternative, because the efficiency of a piezoelectric motor is independent of size.

A piezoelectric motor converts the small displacements of a piezoelectric actuator, or actuators, into linear or rotational movement. Relative to electric motors, piezoelectric motors are more expensive, but they offer several significant advantages, in addition to scalability to very small sizes without sacrificing efficiency. Additional advantages of ultrasonic piezoelectric motors include low speed / high torque (as opposed to high speed / low torque, characteristic of electromagnetic motors), rapid response, a high power /

Figure 4.14 Flextensional transducers

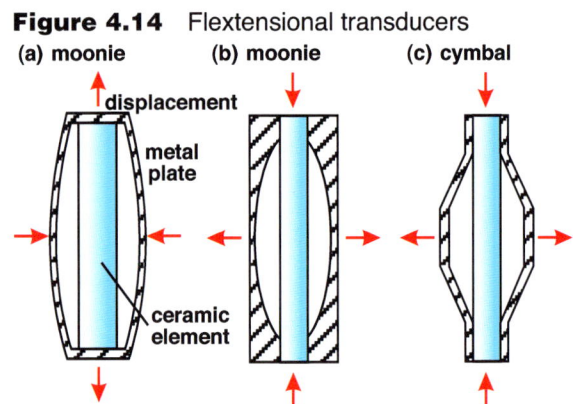

(a) moonie   (b) moonie   (c) cymbal

displacement
metal plate
ceramic element

weight ratio, and high efficiency, and they do not generate electromagnetic fields. Unlike electric motors, which require a finite time to achieve full running torque, piezoelectric motors generate high torque at start-up. They are compact, light weight, and quiet. On the other hand, they require a high frequency power supply, friction drive makes them less durable than electromagnetic motors, and torque falls off as speed is increased.

A piezoelectric motor is an ultrasonic, resonating displacement device—displacement is created by an alternating strain induced by an alternating current field at the mechanical resonance frequency of the device (4). The basic components of an ultrasonic piezoelectric motor are shown in Figure 4.16. The high frequency power supply induces strain in the piezoelectric ceramic driving element, and hence vibration in the elastic vibrating element. The friction coat transfers the vibrations to the elastic sliding element. Very high speed motion, due to the high frequency, is characteristic of ultrasonic piezoelectric motors. Consequently, an ultrasonic motor must be constructed from a very hard piezoelectric ceramic with a high mechanical quality factor, to minimize heat generation and maximize displacement.

Three general types of piezoelectric motors currently are used in various applications: linear, axial push, and travelling wave or standing wave motors (Figure 4.17). A linear motor, or inchworm motor, consists of three actuators, two acting as brakes, the third creating the movement, in a three-step cycle. In step 1, brake 1 is activated, causing the movement actuator to elongate; in step 2, brake 2 is activated; in step 3, brake 1 is deactivated, allowing the movement actuator to contract, but in a new position. The speed of this movement is regulated via the step amplitude and frequency. The wedgeworm actuator (Figure 4.17b) is a simpler and more economical variant on the inchworm design.

In an axial push motor, a stator constructed of ceramic disks and metal cylinders is brought into mechanical resonance. The axial movement is transformed into elliptical movement against a rotor, putting the rotor into motion.

Figure 4.15 Applications for piezoelectric ceramic actuators

(a) optical instruments: adjusting lenses or mirrors

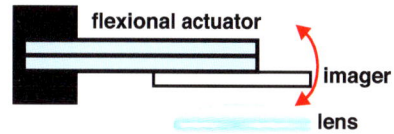

(b) valves

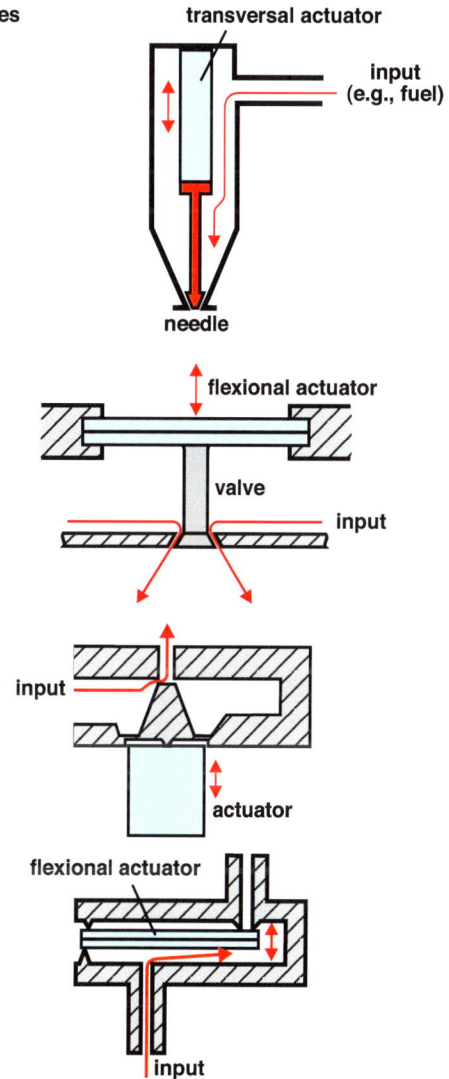

(c) machine tools: precise adjustment

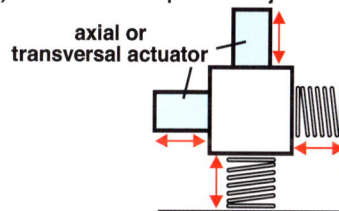

Figure 4.16 Basic components of an ultrasonic piezoelectric motor

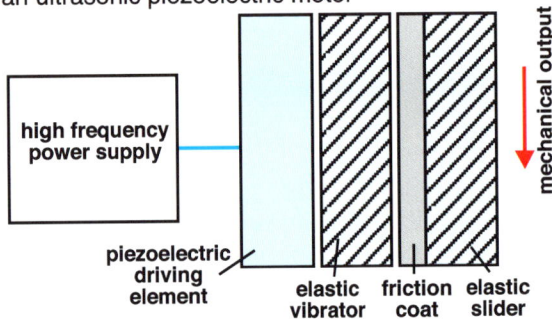

high frequency power supply

piezoelectric driving element

elastic vibrator friction coat elastic slider

mechanical output

In a travelling wave motor, a ring of specially polarized ceramic elements is bonded to a stator ring. Alternating voltage is adjusted to produce resonance and generate a wave. The wave is set into clockwise or counter-clockwise rotation, and the upper part of the stator ring makes elliptical movements against the surface of the rotor, which is specially coated to improve friction while minimizing wear. A traveling wave motor has several valuable characteristics: high torque, low rotation speed (rpm), fast responses, a very flat profile, and no need for a brake. On the other hand, stator rings are expensive to produce and the circuitry needed to operate the motor is complicated, so a travelling wave motor is used only where its unusual features are of particular value.

Figure 4.18 shows two views of the Sashida traveling wave motor, a design that is in commercial production. The thin profile, combined with energy-saving operation, has made this motor a practical autofocusing device for camera lenses. Silent drive makes it an appealing autofocusing mechanism for microphone-equipped video-cameras. The motor can be driven in either direction by exchanging the voltage inputs. Variations of the Sashida motor have been produced in sizes down to 10 mm diameter by 4.5 mm high (Figure 4.19).

The piezoelectric element in a traveling wave motor is a fairly complex construction. Relative to traveling wave motors, standing wave motors are less complicated: they require only one, uniformly poled, piezoelectric ceramic element, simpler electrical wiring, and one power supply. A rotary-type standing wave motor of approximately 3 mm diameter has been constructed of a piezoelectric ceramic ring and metal endcaps cut to form four arms ("windmill" shape) and bonded together to generate up and down and torsional coupled vibration (Figure 4.20). Such a simple motor is easy and inexpensive to produce, making it practical for use in disposable applications.

Heat is an important factor to consider when designing a piezoelectric motor. Heat generated in the motor can elevate the temperature sufficiently to depolarize the ceramic element. The piezoelectric element, therefore, should be prepared from a ceramic with a high Curie point and a high mechanical quality factor (6). For PZT piezoelectric ceramic elements, the mechanical

Figure 4.17 Piezoelectric motors

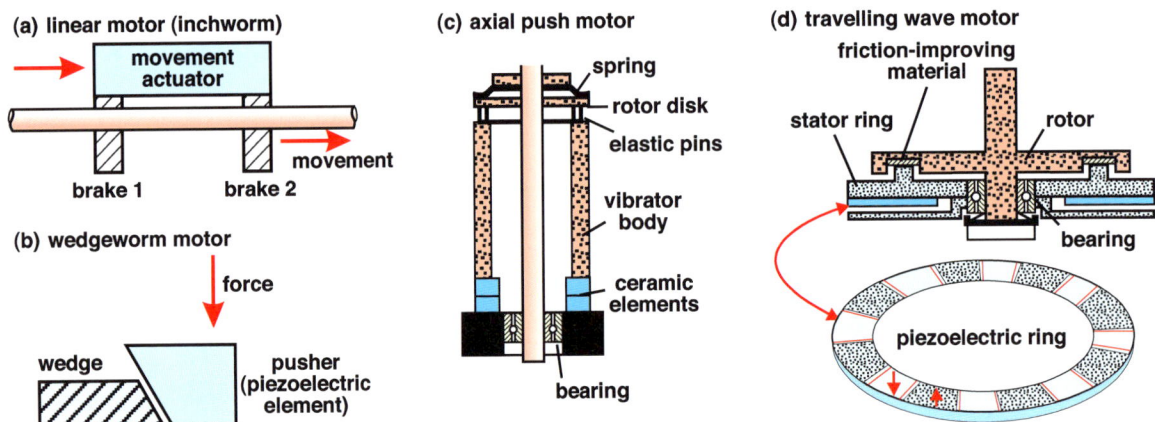

(a) linear motor (inchworm)

movement actuator

brake 1 brake 2 movement

(b) wedgeworm motor

force

wedge pusher (piezoelectric element)

(c) axial push motor

spring
rotor disk
elastic pins
vibrator body
ceramic elements
bearing

(d) travelling wave motor

friction-improving material
stator ring rotor
bearing
piezoelectric ring

quality factor at antiresonance frequencies is larger than the mechanical quality factor at resonance frequencies, and the temperature rise is lower over the entire vibration velocity range (7).* Thus, use of the antiresonance mode appears to offer advantages, relative to the conventional use of the resonance mode, particularly for high power applications, such as ultrasonic motors (7).

Figure 4.18 Sashida motor

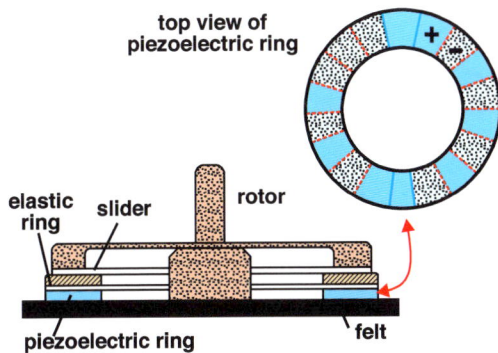

Figure 4.19 Compact traveling wave motor

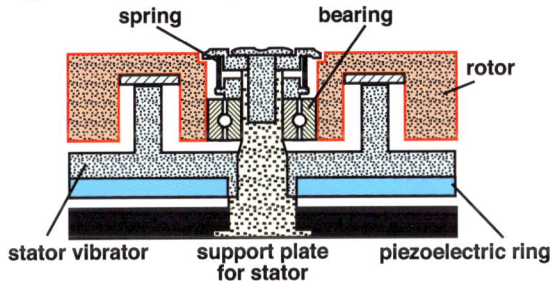

Figure 4.20 Standing wave motor

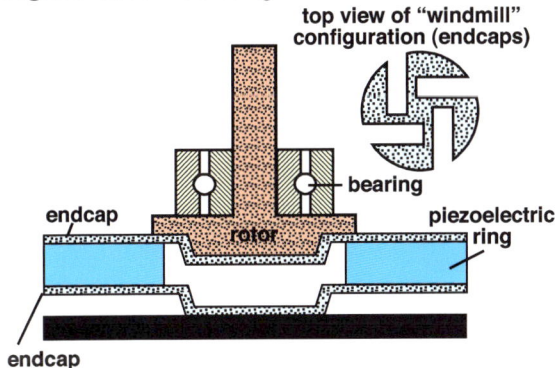

Piezoelectric Pumps

Alternate lengthening and contraction of a piezoelectric actuator can operate an input-output valve, moving small volumes of a gas or liquid. If the dynamic properties of the fluid being pumped are sufficient to control the flow, valving is unnecessary. A ceramic disk or cylinder can move a small volume (0.001 - 0.01 mm^3) at high pressure; a flexional actuator composed of a coin-shaped bilaminar element (two ceramic layers bonded together) or a single ceramic layer bonded to a metal membrane can move volumes up to several mm^3, at much lower pressure. A radially polarized tube also can act as a pump, because the volume of the bore decreases / increases when voltage is applied / removed.

Figure 4.21a shows the basic construction of a pump incorporating a piezoelectric ceramic disk. To prevent depolarization of the ceramic element, the electric field is applied in the direction in which the element is polarized. If the field is supplied by AC voltage, DC-bias voltage can be added. The change in the thickness of the disk can be determined from:

Equation 4.17
$$\Delta h = d_{33}V$$

or

Equation 4.18
$$\Delta h = d_{33}hE$$

or

Equation 4.19
$$\Delta h = s_{33}hT$$
where
h = height (thickness) of ceramic element
d_{33} = piezoelectric charge constant
V = voltage
E = electric field
s_{33} = elastic compliance
T = stress on element**
** applied force / surface area of ceramic element

* Vibration velocity $v_0 = (\sqrt{2})\,(\pi f u_{max})$, where f = resonance or antiresonance frequency, u_{max} = maximum vibration amplitude of piezoelectric element.

The amount of fluid displaced by each extension / contraction cycle is determined by the change in the volume of the disk, Δvol:

Equation 4.20

$$\Delta vol = d_{33}VA$$

where

d_{33} = piezoelectric charge constant
V = voltage
A = surface area of ceramic element

The pressure required to block the displacement, p_B, is:

Equation 4.21

$$p_B = (d_{33}V) / (s_{33}h)$$

where

d_{33} = piezoelectric charge constant
V = voltage
s_{33} = elastic compliance
h = height (thickness) of ceramic element

p_B also is the pressure exerted by the fluid being pumped.

Figure 4.21b shows how a flexional ceramic element can act as a pump. Fluid displacement can be estimated by determining the deflection at the center of the disk, e.g., for APC 850 or equivalent Navy Type II ceramic:

Equation 4.22

$$d_d = 10^{-10} (D_\phi^2 / h^2) V$$

Equation 4.23

$$\Delta vol = (0.5) (\pi D_\phi^2 / 4) (d_d)$$

where

d_d = deflection of ceramic element (m)
D_ϕ = diameter of ceramic element
h = height (thickness) of ceramic element
V = voltage
Δvol = volume of displacement (m³)

An approximate value for the blocking pressure, p_B, can be calculated from Equation 4.24. As mentioned earlier, pressure will be much lower than for a single element disk or tube.

Figure 4.21
Piezoelectric actuators as pumps

(a) single element ceramic disk

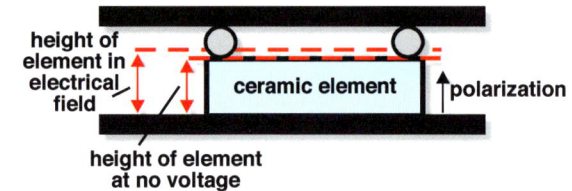

(b) flexional disk

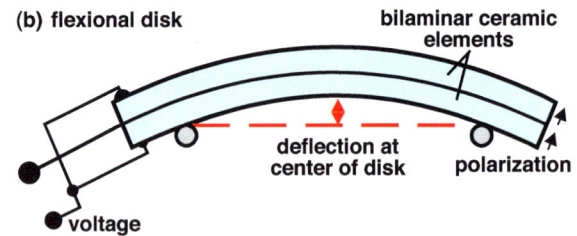

Equation 4.24

$$p_B \cong (8) (h / D_\phi^3) (V)$$

where

p_B = blocking pressure (Pa)
h = height (thickness) of ceramic element
D_ϕ = diameter of ceramic element
V = voltage

Applications for Flexional Actuators

Flexional actuators, and Stripe actuators in particular, are well suited to applications that demand sensitivity and a large response. They have been used to control needle movements in textile weaving machines, in switching applications, including touch-probe switches, in Braille machines, and as small-volume pumping devices. A Stripe actuator can open and close a valve, position a lens or mirror in an optical device, or maintain an electrical contact. Alternatively, Stripe actuators also are very useful in sensing applications.

Rotary-type electromagnetic cooling fans in computers and other electronic devices produce unwanted electromagnetic noise. In addition to eliminating noise, a fan constructed from a piezoelectric bilaminar flexing element is simpler, consumes little power (less than 100 mW), and has a long operating lifetime (estimate of approximately 25,000 hours) (8).

References

1. Wang, Q.-M. and L.E. Cross, *A piezoelectric pseudoshear multilayer actuator*
   Vol. III/Appendix 18 in *Acoustic Transduction — Materials and Devices*
   Annual Report (1 January 1998 to 31 December 1998) Office of Naval Research, Contract No: N00014-96-1-1173 (Apr. 1999).
2. Fraden, J. *Handbook of Modern Sensors* (2nd Ed.)
   American Institute of Physics Press, Woodbury, NY (1997).
3. Yao, K., W. Zhu, K. Uchino, Z. Zhang, and L.C. Lim, *Design and Fabrication of a High Performance Multilayer Piezoelectric Actuator with Bending Deformation*
   Vol. III/Appendix 51 in *Acoustic Transduction — Materials and Devices*
   Annual Report (1 January 1999 to 31 December 1999) Office of Naval Research, Contract No: N00014-96-1-1173 (June 2000).
4. Uchino, K. and B. Koc, *Compact Piezoelectric Ultrasonic Motors*
   Vol. IV/Appendix 62 in *Acoustic Transduction — Materials and Devices*
   Annual Report (1 January 1999 to 31 December 1999) Office of Naval Research, Contract No: N00014-96-1-1173 (June 2000).
5. Uchino, K., *Ferroelectric Devices*
   Marcel Dekker, New York (2000).
6. Uchino, K., *Piezoelectric Actuators and Ultrasonic Motors*
   Kluwer Academic Publishers, Boston / Dordrecht (Netherlands) / London (1997).
7. Uchino, K. and S. Hirose, *Loss Mechanisms in Piezoelectrics — How to Measure Different Losses Separately*
   Vol. IV/Appendix 69 in *Acoustic Transduction — Materials and Devices*
   Annual Report (1 January 1999 to 31 December 1999) Office of Naval Research, Contract No: N00014-96-1-1173 (June 2000).
8. Yoo, J.H., J.I. Hong, and W. Cao, *Piezoelectric Ceramic Bimorph Coupled to Thin Metal Plate as Cooling Fan for Electronic Devices*
   Vol. III/Appendix 55 in *Acoustic Transduction — Materials and Devices*
   Annual Report (1 January 1999 to 31 December 1999) Office of Naval Research, Contract No: N00014-96-1-1173 (June 2000).

Request annual reports to Office of Naval Research from:

Office of Naval Research
Regional Office Chicago
536 S. Clark Street
Room 208
Chicago, IL 60605-1588.

transducers

Piezoelectric transducers convert electrical energy into vibrational mechanical energy, usually sound or ultrasound, that is used to perform a task. Transducers that generate audible sounds are compact, simple, and highly reliable, and use minimal energy to produce a high level of sound. Ultrasonic signals generated by piezoelectric transducers are employed to measure distances in air, water, or other fluid media, to determine flow rates, or to monitor fluid levels. Ultrasonic vibrations are used for cleaning, to atomize liquids, to drill or mill ceramics or other difficult materials, to weld plastics, or for medical diagnostics. Because the piezoelectric effect is reversible, a transducer can both generate an ultrasound signal from electrical energy and convert incoming sound into an electrical signal, but the two functions often are separated to optimize the performance of each task.

large containers (e.g., grain silos), proximity-warning devices, and intruder alarms. The operating range for these *impulse-echo* devices is determined by the operating frequency and the power of the transducer. The latter factor, power output, is a function of the mechanical characteristics and thermal constraints of the device. Operating frequency requires careful consideration because signal reflection and absorption are frequency dependent. For long range applications, low frequency signals are subject to much less signal damping than high frequency signals. For the best signal directivity and object resolution, however, the frequency should be as high as possible. For accurate distance measurements, especially at short ranges, the excitation pulses must be short.

Narrowing the signal bandwidth extends the range of an impulse-echo device, reduces the input power

The working component in an audible sound transducer usually is a thin disc of piezoelectric ceramic bonded to a similarly thin metal membrane. When a voltage is applied to the ceramic disc, the disc deforms, causing the metal membrane to bend. When an alternating voltage is applied the ceramic / metal element vibrates at the frequency of the applied voltage, producing audible sound. (The frequency required to vibrate the ceramic disc alone would be too high to produce audible sound.)

An ultrasonic transducer consists of a piezoelectric ceramic element bonded to a thin, flexible metal diaphragm and mounted in an open or sealed housing. These devices offer high sensitivity, superb sound pressure level, and stable electrical and mechanical characteristics.

Apparatus in which piezoelectric transducers are used to measure distances in air include level detectors for required, and minimizes interference and reflections from objects outside the signal path. If the signal is too narrow, however, atmospheric conditions can divert the signal from its intended path, particularly at longer distances. Also, a broader band will exhibit less noise.

Transducers

A piezoelectric transducer converts an electrical voltage or charge into vibrational mechanical energy which, in turn, is used to perform a task (ultrasonic cleaning, measuring distances, etc.). In signal transmitting and receiving applications, a piezoelectric transducer will, through the reverse piezoelectric effect, convert an incoming audible sound or ultrasound signal into an electrical signal. Transducers can be categorized by the nature of the vibrational energy they produce and the application to which the energy is used. The general categories are:

- devices that produce audible sound signals (frequencies below 20 kHz)
- devices that produce ultrasonic vibrations for transfer to a liquid or solid medium
- devices that produce ultrasonic signals (transmitter) which are received by a second transducer (receiver)

  or

  devices that alternate between transmitting an ultrasonic signal and receiving a reflected signal (transmitter / receiver)

Audible sound transducers are the core components of buzzers, telephone microphones, and high-frequency loudspeakers. Transducers that transfer ultrasonic frequencies to a liquid or solid medium are used in ultrasonic cleaning equipment, liquid-pulverizing devices, including atomizers and air humidifiers, equipment for welding plastics, and ultrasonic therapy equipment. Devices that transmit / receive ultrasonic signals are used for measuring distances in air or in liquids, for imaging, for determining flow rates, and in fluid level sensors and signal delay lines. Among the transducers that measure distances are intruder alarms, proximity-warning devices (parking aids), detectors for monitoring fill levels

in large storage containers, and echo sounding equipment (e.g., sonar, fish detectors). Imaging applications include medical and industrial diagnostics (e.g., fetal heart monitors, apparatus for structural integrity analysis). When delivery of an electrical signal must be delayed, a piezoelectric delay line offers significant advantages relative to an electronics-based delay system.

Audible Sound Transducers

Simple, compact, and reliable, piezoelectric audible sound transducers—also called tone generators, or buzzers—can deliver a high sound output from a small (milliwatt) energy input. Emitted sounds range from soft hums to strident alarms. These devices are well suited for use in portable, battery powered equipment, and are employed in a wide variety of products, including timers, smoke alarms, games, telephone ringers, metal detectors, watches, automobile alarms, and many others.

The working component in most audible sound transducers is a thin disk of piezoelectric ceramic bonded to a similarly thin metal diaphragm (Figure 5.1). When a voltage is applied to the ceramic disk, the disk deforms, causing the metal diaphragm to bend. When a recurring voltage is applied

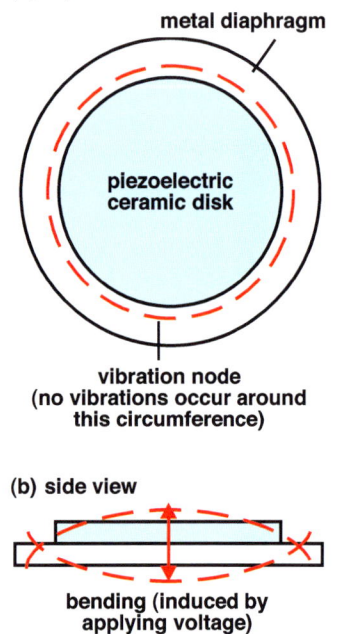

Figure 5.1 Sound-producing element in an audible sound transducer

(a) top view

metal diaphragm

piezoelectric ceramic disk

vibration node
(no vibrations occur around this circumference)

(b) side view

bending (induced by applying voltage)

the ceramic / metal bending element vibrates at the frequency of the applied voltage, and produces an audible sound. If the mechanical resonance frequency of the ceramic / metal element and the frequency of the applied electrical signal are matched, the amplitude of the vibrations will be greatest, and sound output will be maximum. (The resonance frequency of the ceramic element alone is too high to produce audible sound, hence the need for the metal diaphragm.)

The resonance frequency, f_r, for the ceramic disk / metal diaphragm bending element in an audible sound transducer can be determined from Equation 5.1:

Equation 5.1

$$f_r = (0.412h / r_{disk}^2) (\sqrt{(Y / \rho[1 - \sigma^{E2}])})$$

where

h = height (thickness) of ceramic / metal bending element

r_{disk} = radius of ceramic / metal bending element

Y = Young's modulus*

ρ = density of ceramic material (kg / m^3)

σ^E = Poisson's ratio**

* ratio of stress applied to strain developed; Y = (force / area of element) / (change in height of element / original height of element)

** compressibility; $\sigma^E = s^E_{12} / s^E_{11}$ where s^E = elastic compliance (transverse contraction strain / longitudinal expansion strain) in a constant electric field.

If the sound-producing ceramic / metal bending element is assumed to be a homogeneous unit, the resonance frequency is proportional to the overall height of the ceramic disk and metal diaphragm (h) and is inversely proportional to the square of the radius of the bending element (r_{disk}):

Equation 5.2

$$f_r = (\alpha) (h / r_{disk}^2)$$

where

$$\alpha = 0.412 (\sqrt{(Y / \rho[1 - \sigma^{E2}])})$$

For any perovskite piezoelectric ceramic σ^E is approximately 0.3 (1), therefore α becomes: $0.412 (\sqrt{(Y / 0.91\rho)})$

Figure 5.2 shows an electrical circuit equivalent to an audible sound transducer. R_m, L_m, and C_m express the mechanical resonance of the sound-producing mechanism, f_r:

Figure 5.2 Electrical circuit equivalent to an audible sound transducer

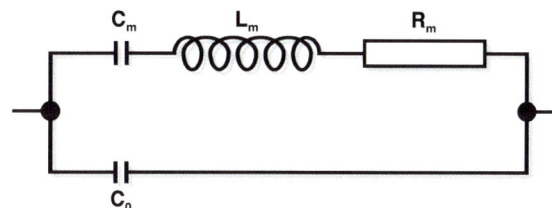

C_0 = (capacitance of ceramic element below resonance frequency) - (c_m)
C_m = capacitance of mechanical circuit
L_m = inductance of mechanical circuit
R_m = resistance caused by mechanical losses

Equation 5.3

$$f_r = 1 / (2\pi) (\sqrt{L_m C_m})$$

where

L_m = inductance of mechanical circuit

C_m = capacitance of mechanical circuit

The manner in which a buzzer is constructed affects the resonance frequency and the level of sound the device emits, the extent of unit-to-unit variation in production, the conditions under which the device can be used, and the cost of the device. Figure 5.3 shows three alternative possibilities for buzzer design. In the nodal support mounting, the flexing element is clamped into the housing around the nodal ring - a ring approximately 2/3 the distance from the center of the element to the periphery of the element, around which there is no vibration when the mounted element is stimulated. This mounting option minimizes the mechanical restriction

Figure 5.3 Audible sound transducers

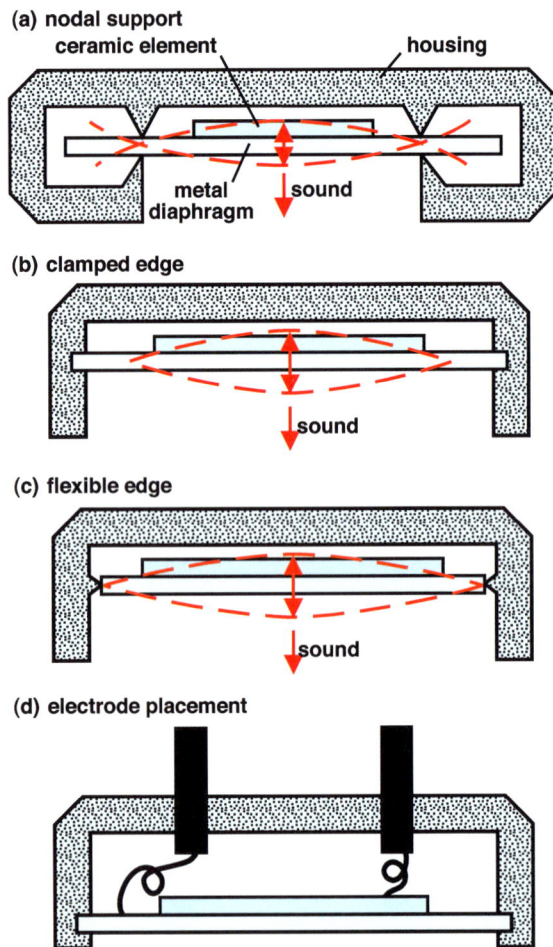

(a) nodal support

ceramic element — housing — metal diaphragm — sound

(b) clamped edge

sound

(c) flexible edge

sound

(d) electrode placement

on the movement of the sound-producing mechanism, so the amplitude of the vibrations is highest. On the other hand, signal originating from the outer part of the element, beyond the nodal ring, will be in antiphase with signal from the central part of the element, potentially causing interference and reducing the sound output. To prevent interference, the housing in which the buzzer is mounted must absorb all output from the outer part of the element. Assuming the potential interference is nullified, nodal support mounting is the best choice for self-driven, high output applications.

In the clamped edge mounting, the flexing element is clamped to the housing around its periphery. This allows the entire surface of the element, except a narrow band around the pe-riphery, to vibrate in phase. Under equivalent conditions, the vibration frequency for a flexing element in a clamped-edge mounting will be approximately the same as would be exhibited by the central part of the flexing element if the element were mounted in a nodal support mounting. Relative to nodal support construction, however, greater interaction between the sound-producing element and the housing reduces the mechanical quality factor of the element and, consequently, the amplitude of the vibrations is lower. Furthermore, unit to unit variations in clamping can affect the uniformity of product performance, and a more substantial housing generally is needed to retain the element.

Flexible edge mounting is similar to clamped edge mounting, except that in the flexible edge design, the flexing element is not rigidly clamped in the housing, but instead is restrained in a flexible material, such as a rubber. The pliant restraint allows angular movements at the periphery of the flexing element, and this ensures excellent signal characteristics. The mechanical resonance frequency for a flexing element in a flexible edge mounting is half that for the same element in a nodal support mounting or a clamped edge mounting. Similar to the constraints on clamped edge mounting, however, careful construction of the device is critical to its effective performance. Flexible edge mounting is used to the most advantage by incorporating it into a system driven by external drive circuitry.

A fourth alternative—supporting the disk at its center—is possible, but the design is difficult to execute, and sound pressures will be lowest with this option.

Helmholtz Resonators

Whether the flexing element is mounted in a nodal support mounting, a clamped edge mounting, or a flexible edge mounting, the acoustic impedance of the flexing element must be matched to that of the surrounding air, or energy transfer will be inefficient, and the sound level will be unsatisfactory. To ensure an excellent sound output, the housing dimensions can be designed to match the acoustic impedance of the air cavity in the housing to the impedance of the flexing element (Figure 5.4). The concept is roughly analogous to putting a music box movement into a wooden or plastic resonating container. A device of this design is a Helmholtz resonator. When a Helmholtz resonator and the flexing element it houses have approximately the same resonance frequency, the vibrations produced by the flexing element and by the housing will be coupled, and the transducer will produce a strong sound output over a broader frequency range (Figure 5.5). The resonance frequency, f_r, for the air cavity of a Helmholtz resonator can be determined from:

Equation 5.4

$$f_r = (v_{air} / 2\pi) \, (\sqrt{[(4r_{hole}^2) / (D_\varnothing^2 h_{cavity})(h_{hole} + 1.3r_{hole})]})$$

where

v_{air} = velocity of sound in air (~344 m/s)*
r_{hole} = radius of hole in housing
$D_\varnothing$ = diameter of element-supporting portion of housing
h_{cavity} = height of air cavity in housing
h_{hole} = height of hole in housing
  * At 24°C; varies with temperature.

Figure 5.4 Helmholtz resonator incorporated into a sound transducer

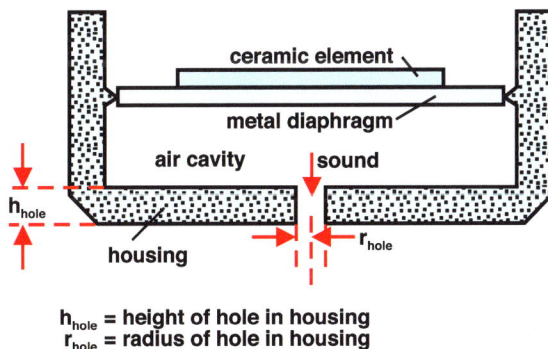

h_{hole} = height of hole in housing
r_{hole} = radius of hole in housing

Figure 5.5 Bandwidth improved by coupling resonances of flexing element and housing

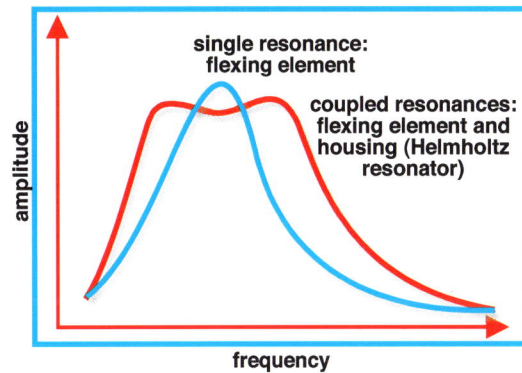

Resonance in a Helmholtz resonator is equivalent to an electrical circuit: the air in the housing absorbs or releases potential energy during pressure variations created by the flexing element (capacitance), and air moving through the holes in the housing absorbs kinetic energy (inductance). The resonance frequency, coupling between the Helmholtz resonator and the flexing element, and damping of the flexing element and resonator determine the specific characteristics of signal frequency.

It is evident that construction dimensions are critical for a Helmholtz resonator. In addition, however, the construction must be airtight and must minimize damping. Because this design affords a broad frequency range, multi-tone buzzers, such as those in telephone equipment, are mounted in Helmholtz resonators.

The flexing element in a piezoelectric buzzer can be driven by square waves, pulsed waves, or sine waves. Square waves or pulsed waves ensure the highest acoustic output, but harmonics also will be highest. A capacitor installed in the system in parallel with the buzzer will reduce the harmonics. If sine waves are used, the element will operate at a frequency lower than its resonance frequency, because of the time lag between full deflections, relative to square waves, and with a lower sound pressure level (SPL), because energy is lost between full deflections.

Sound Pressure

The sound pressure generated by an audible sound-producing device is a function of the loudness and the frequency of the sound. The value for the sound pressure is the ratio of the measured pressure produced by the device, relative to a pressure of 20 μPa, the lowest sound pressure that can be discerned by an average person. This calculated ratio is converted to a logarithmic value and the pressure is expressed in decibels (dB):

Equation 5.5

$$dB = 20 \log (\text{measured pressure from device} / 20 \text{ μPa})$$

By such calculations a sound at the threshold of human hearing, a 1:1 ratio relative to the standard denominator, is 0 dB. The point at which sound pressure begins to be painful, approximately 60 Pa, is a ratio of 3,000,000:1 relative to the standard denominator, or 130 dB. As the sound travels away from its source it enters continuously greater air volume, with corresponding reduction in pressure, until the value falls below the threshold of human hearing.

Electrical Connections / Drive Circuits

A piezoelectric buzzer comprised of a ceramic element and a metal diaphragm can have either two or three electrical leads. In a two terminal pattern the metal diaphragm and the ceramic element are the two electrode faces (Figure 5.6). Piezoelectric buzzers with two leads have capacitive impedance, and drive circuits for these elements are designed by treating the transducer as a capacitor. A three terminal pattern, or feedback pattern, is a split electrode design: a signal applied between electrodes 1 and 2 produces a phase-shifted signal between electrodes 1 and 3, imitating a piezoelectric transformer (see **Transformers** in **Transmitting Ultrasonic Signals**, this chapter). The phase-shifted signal can be used as the feedback component in a simple three-terminal oscillation circuit, which will operate automatically at the resonance frequency of the ceramic element. A buzzer with a two terminal pattern is operated via an external drive circuit; a buzzer with three terminals is operated via a self-drive circuit (Figure 5.7).

If the application for a buzzer does not require a rigidly fixed signal frequency, the buzzer can be made to vibrate at its resonance frequency by incorporating it into a drive circuit with a feedback loop (Figure 5.8). The bandwidth will be narrow and the self-stabilizing system will allow the frequency to vary somewhat, but a high mechanical quality factor will ensure a low energy consumption and make the device compatible with battery-drive operation. A feedback electrode simplifies the drive circuitry (Figure 5.8) but, alternatively, the circuit can be based on a multivibrator. The frequency of the multivibrator is, in turn, controlled by the flexing element. These circuits allow leeway in manufacturing tolerances for the buzzer.

If necessary, a buzzer can be made to operate at a fixed frequency, dictated by the circuit, regardless of unit to unit differences in manufacturing or variations in operating temperature. To circumvent the effects of unit to unit variation in manufacturing, this system requires a transducer with a larger bandwidth. A Helmholtz resonator can be used to ensure a suitable signal at the specified frequency.

A flexing element can be made to vibrate at its resonance frequency by incorporating an inductor in the circuitry, in parallel with the flexing element (Figure 5.9). Switching off the current through the inductor will induce voltages much higher than the supply voltage, and will cause the element to vibrate at resonance frequency. Like a drive circuit with a feedback loop, a drive circuit with a parallel inductor negates the influence of unit to unit variations in transducer construction.

High amplitude square waves also will vibrate the flexing element at its resonance frequency, but the broad frequency spectrum of square waves will generate harmonics by the buzzer.

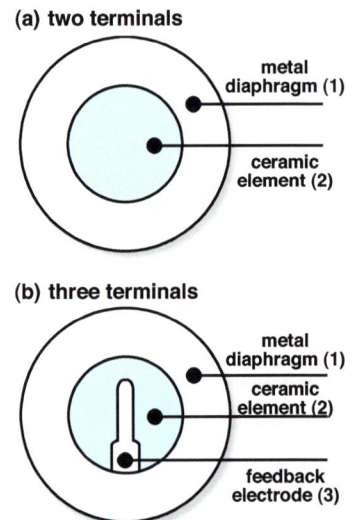

Figure 5.6 Electrical connection patterns for piezoelectric buzzers

(a) two terminals

metal diaphragm (1)

ceramic element (2)

(b) three terminals

metal diaphragm (1)

ceramic element (2)

feedback electrode (3)

Figure 5.7 Piezoelectric buzzers with self-drive and external drive electrical circuits

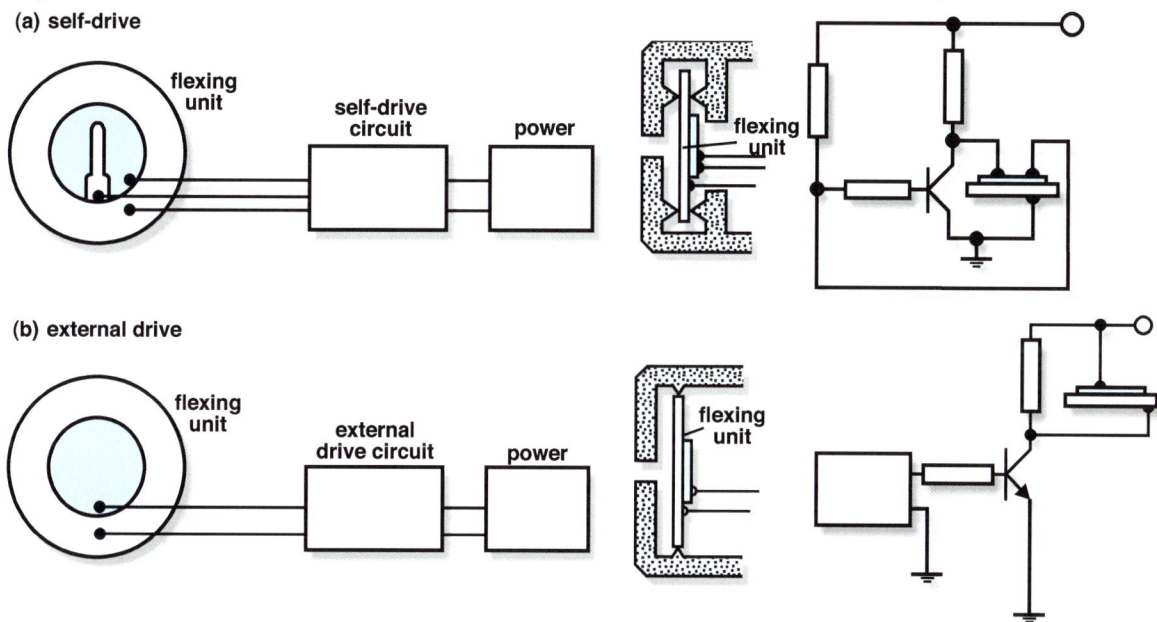

(a) self-drive

(b) external drive

Drive circuits are classified as either amplification circuits or oscillation circuits. Amplification circuits are stable circuits that amplify the input signal and supply it to the transducer. In oscillation circuits the electrical signal varies over time (i.e., the circuit is unstable). The transducer is an integral part of the oscillation circuit design, along with other active elements.

An amplification circuit fits one of three types: load resistance, load inductance, or complementary. Load resistance circuits (Figure 5.10) are the simplest amplification circuits, but they cannot provide a loud sound signal. On the other hand, two-step amplification can improve the sound output (Figure 5.11). Load inductance (Figure 5.12) allows high voltages to be applied to the transducer: the smaller the inductance, the higher the peak voltage allowed. Current consumption is greater than for load resistance circuits or complementary circuits, however, and higher levels of spurious oscillation reduce tone quality. Complementary circuits (Figure 5.13) are highly efficient, and reduce current consumption, but the sound pressure level can be relatively low.

There are five types of oscillation circuits: three-terminal circuits, integrated circuits (IC), RC (resistor-capacitor) circuits, multivibrator circuits,

and blocking circuits. In a three-terminal circuit the feedback connection on the transducer is used to make the circuit oscillate, eliminating the need for inductors and capacitors. The simplified system can produce a variety of sound signals, from sine waves

Figure 5.8 Drive circuit with feedback loop

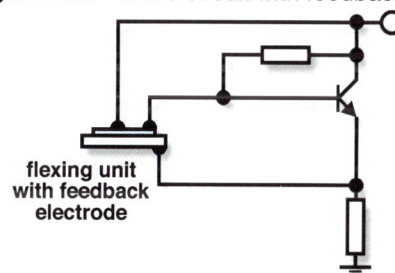

Figure 5.9 Drive circuit with parallel inductor

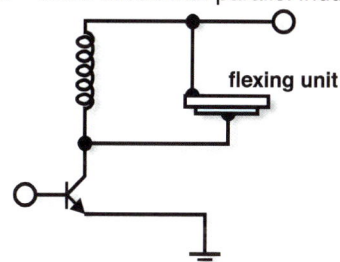

Figure 5.10 Amplification-type drive circuit: load resistance

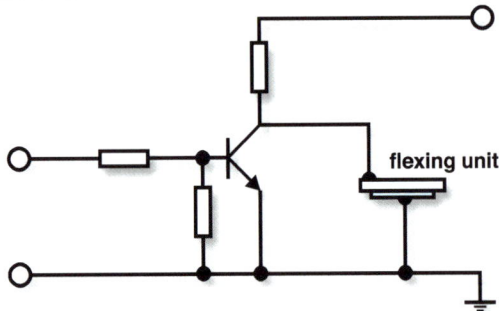

flexing unit

Figure 5.11 Two step amplification in load resistance drive circuit

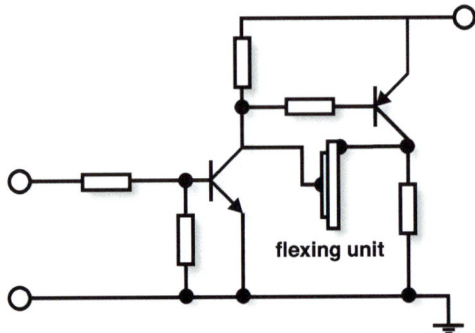

flexing unit

Figure 5.12 Amplification-type drive circuit: load inductance

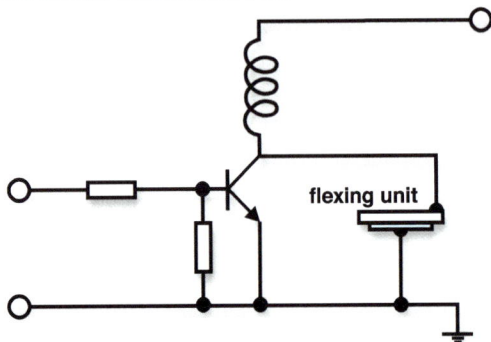

flexing unit

Figure 5.13 Amplification-type drive circuit: complementary

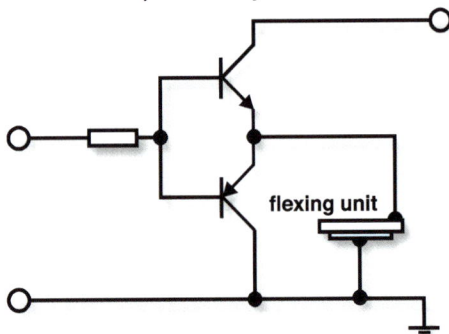

flexing unit

to trapezoidal waves. Integrated circuits (Figure 5.14) incorporate logic circuits (complementary metal oxide-silicon inverters (CMOS), CMOS NAND gates, etc.) to produce a variety of sounds from the same transducer. In RC oscillation circuits (Figure 5.15) the operating frequency is defined by resistors and capacitors. RC circuits are capable of generating a sine wave when the high frequency component is small, and thus produce good tone quality. Transistors with higher gain (h_{fe}) values improve the stability of the oscillation. Sound pressure is low. In a multivibrator circuit (Figure 5.16) the transducer is connected between the collectors of two transistors of a multivibrator that generates a stable square wave. Peak-to-peak voltage (V_{pp}) is double the supply voltage (V_{cc}). A blocking circuit (Figure 5.17) is the most effective circuit when loud sound pressure levels are desired and the supply voltage is low. The duty factor will be large and the spurious oscillation will increase if the inductance of the transformer is small, much the same as for a load inductance (amplification type) circuit.

A booster coil can be used to compensate for loss of sound pressure (Figure 5.18).

Figure 5.14 Oscillation-type drive circuits: integrated circuits

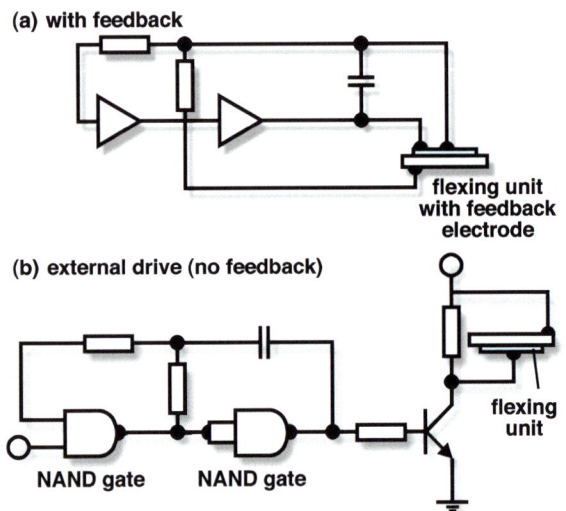

(a) with feedback

flexing unit with feedback electrode

(b) external drive (no feedback)

NAND gate   NAND gate

flexing unit

Figure 5.15 Oscillation-type drive circuit: RC

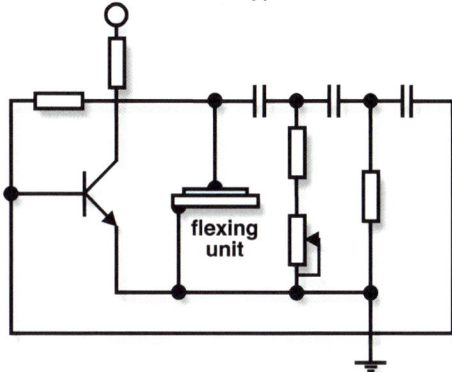

flexing
unit

Figure 5.16 Oscillation-type
drive circuit: multivibrator

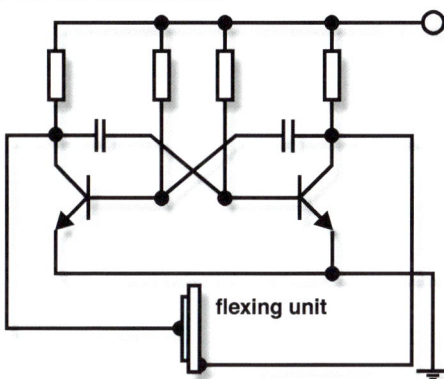

flexing unit

Figure 5.17 Oscillation-type
drive circuit: blocking

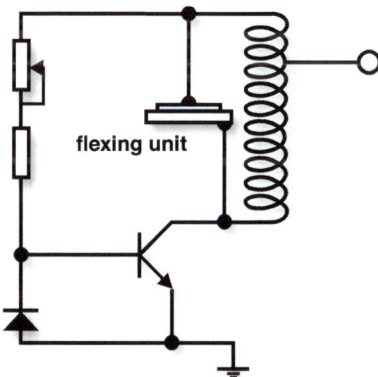

flexing unit

Figure 5.18 Drive circuit with booster coil

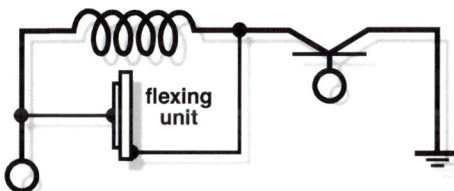

flexing
unit

Generating Ultrasonic Vibrations in Liquids or Solids

Ultrasonic Cleaning: Disk Transducers

A ceramic disk or a ceramic disk / metal disk combination, bonded to a water tank, makes a simple but effective ultrasonic cleaning apparatus (Figure 5.19). There are considerable performance differences between a ceramic disk transducer and a ceramic disk / metal disk transducer, however. A ceramic disk bonded directly to the water tank vibrates in the radial mode and exhibits an impedance of several hundred ohms. In contrast, positioning a metal disk between the water tank and the ceramic disk makes the tank wall an integral part of the transducer. The radial vibrations from the ceramic disk apply shear forces to the tank wall, causing the wall to flex and transfer energy to the liquid in the tank. A cleaning device of this construction will exhibit an impedance of several thousand ohms.

Figure 5.19 Disk
resonator: simple but
effective design for ultrasonic
cleaning

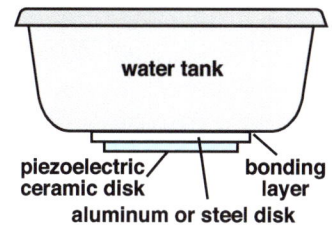

water tank

piezoelectric/
ceramic disk
bonding
layer
aluminum or steel disk

In the design of an ultrasonic cleaning apparatus, the thickness of the cleaning tank wall dictates the thickness of the ceramic disk. Often, 1 mm stainless steel is used to form the tank, in which case a well-matched disk will be approximately 3 mm thick. Because substantial heat will be generated while the cleaner is in operation, the disk should be made from a ceramic with a high Curie point (>300°C), to prevent thermal depolarization. When a metal disk is incorporated into the transducer, it usually is of the same thickness as the ceramic disk, but a larger diameter is required, to give the two disks equal radial mode resonance frequency. Typically a 50 mm ceramic disk combined with an aluminum or steel disk approximately 80 mm in diameter will produce the desired resonance frequency—between 40 kHz and 60 kHz. Aluminum provides a greater electromechanical coupling factor than steel and consequently is the preferred material for the metal disk.

Ultrasonic Cleaning:
Composite Transducers

As is true in many other applications for piezoelectric materials, an assembly of multiple ceramic elements offers considerable performance and production advantages in ultrasonic cleaning applications, relative to a single ceramic element. In order to provide the most efficient operation, simplify manufacturing, and reduce costs, more complex transducers intended for ultrasonic power applications usually are a composite of a piezoelectric ceramic center (multiple thin rings or disks of ceramic, for example), encompassed by metallic end or top and bottom parts (Figure 5.20). Under no liquid load, the mechanical quality factor, Q_m, for a well-designed composite transducer will be greater than the corresponding value for an equivalent single-piece ceramic transducer, and efficient heat conduction by the metallic portions will ensure a lower operating temperature in the ceramic portion of the transducer. The coupling factor, k, will approach that for a single-piece ceramic transducer.

The metallic portions of a composite transducer should have the same acoustic properties and cross-sectional area as the ceramic portion.

Figure 5.20 Composite
transducer for ultrasonic applications

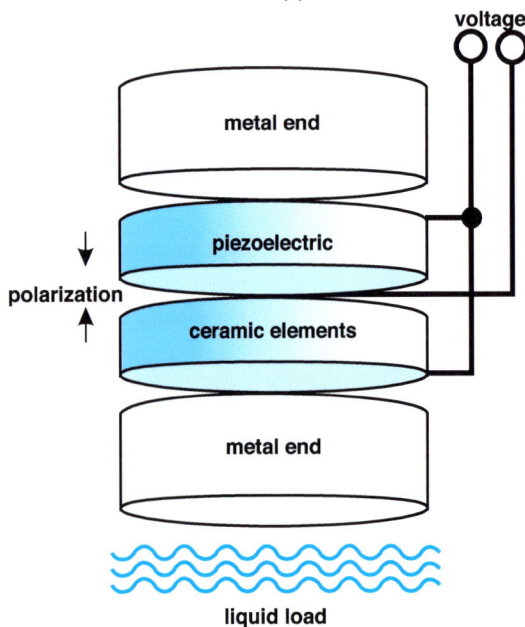

Both metallic parts can be constructed from the same material or combination of materials, or the two parts can be made from materials with divergent properties. Potential construction materials include steel, aluminum, titanium, magnesium, bronze, and brass. Often, only one of the metallic parts is intended for high intensity output.

For maximum energy transfer from the transducer to the solvent in the ultrasonic cleaning tank, a composite ultrasonic transducer usually is a half-wavelength transducer with a resonance frequency of 20 kHz or 40 kHz. The electroacoustic efficiency, η, of a composite ultrasonic transducer has an inverse relationship with the electromechanical coupling factor and the various quality factors of the components:

Equation 5.6

$$\eta \cong 1 - [\,(1 \,/\, (1 + k^2 Q_e Q_l)) - (1 \,/\, (1 + Q_m \,/\, Q_l))\,]$$

where

k = electromechanical coupling factor

Q_e = electrical quality factor

Q_l = acoustic load quality factor

Q_m = mechanical quality factor
   (no load on system)

In Equation 5.6, component $(1 \,/\, (1 + k^2 Q_e Q_l))$ represents the dielectric losses for the system; component $(1 \,/\, (1 + (Q_m \,/\, Q_l))$ represents the mechanical losses. When Q_l is optimized, electroacoustic efficiency is optimized:

Equation 5.7

$$Q_l = (1 \,/\, k)\, (\sqrt{\,(Q_m \,/\, Q_e)\,})$$

Equation 5.8

$$\eta = 1 - (2 \,/\, k\sqrt{(Q_e Q_m)})$$

where

$$k \sqrt{(Q_e Q_m)} \ll 1$$

At high drive frequencies, factors Q_e and Q_m are not constants, but these factors often fall to much smaller values than the corresponding values at low drive frequencies.

Pre-Stressed Composite Ultrasonic Transducers

Seldom will the ceramic component of a composite transducer have adequate tensile strength to withstand the high mechanical stress associated with the power demands for ultrasonic cleaning applications. The tensile strength of the ceramic elements can be supplemented by mechanically pre-stressing the elements along the direction of polarization. Pre-stress is introduced by incorporating a single, large, central bolt or several smaller, peripherally arranged bolts into the design of the transducer (Figure 5.21). The single central bolt design offers slightly higher efficiency than the multiple peripheral bolt design, but manufacturing costs can be higher, assembly can be more difficult and, physically, the transducer will be significantly longer.

For the conditions under which ultrasonic cleaning devices are operated, pre-stress of less than approximately 30 MPa usually is sufficient to protect the ceramic components of the transducer. On the other hand, if the pre-stress is too low, excessive mechanical losses at the ceramic / metallic interfaces can reduce efficiency. Pre-stress can be estimated by using a torque wrench, calibrated against charge, to tighten the bolts. This measuring method is simple, but it is not the most accurate, and consequently it is recommended primarily for production-run transducers for which the variations among corresponding components are, hopefully, minimal. The more accurate way of measuring pre-stress is to measure the charge generated in the ceramic elements under short-circuit conditions. A capacitor connected to the transducer's electrical terminals and to a direct current voltmeter facilitates measurement of the charge as each bolt is tightened.

In ultrasonic cleaning applications, effects of the dimensions and configuration of the water tank, the water load, and the thickness of the bonding layer affixing the transducer to the water tank combine to slightly reduce the frequency of the transducer, and give rise to several additional resonances. Despite these negative factors, however, a well-designed transducer, incorporated in a well-designed circuit, will operate near its resonance frequency.

Cavitation

Cavitation occurs when the vibration of a transducer surface interfacing with a liquid (usually water, but possibly a water / organic solvent mixture in an ultrasonic cleaning application) is sufficient to create a partial vacuum that exceeds the vapor pressure of the liquid, and bubbles form at the vibrating surface. Cavitation is a desirable feature in ultrasonic cleaning or liquid vaporizing applications, but obviously must be avoided in signal transmission applications.

At atmospheric pressure, and with water alone as the interfacing liquid, the threshold for cavitation, p_{C0} (bar), is:

Equation 5.9

$$p_{C0} = (0.00025\,f)^2 + (0.045\,f - 1)$$

for frequencies from kilohertz to several hundred kilohertz (2). If the transducer is submerged, and the vibrating surface is several meters or more below the atmosphere / liquid interface, the threshold for cavitation increases to:

Equation 5.10

$$p_{Ch} = p_{C0} + 0.10h$$

where

p_{Ch} = threshold for cavitation in atmospheres at depth h in meters (bar)

p_{C0} = threshold for cavitation at zero depth (bar)

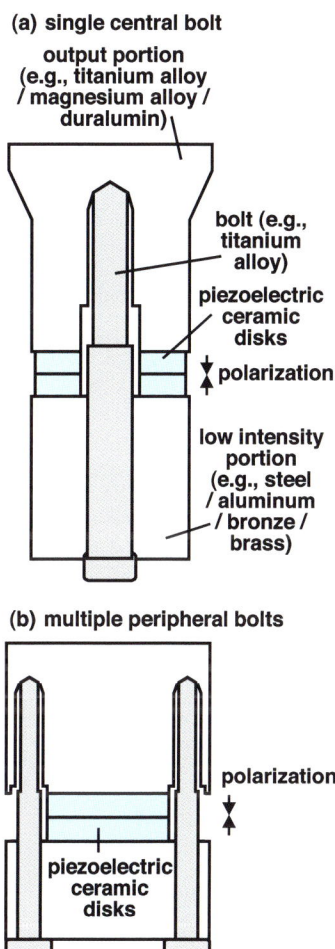

Figure 5.21 Pre-stressed composite transducers

(a) single central bolt

output portion (e.g., titanium alloy / magnesium alloy / duralumin)

bolt (e.g., titanium alloy)

piezoelectric ceramic disks

polarization

low intensity portion (e.g., steel / aluminum / bronze / brass)

(b) multiple peripheral bolts

polarization

piezoelectric ceramic disks

Cavitation (J) can be initiated when the acoustic intensity at the vibrating surface, in W/cm^2, is:

Equation 5.11

$$J = 0.15 \, (p_{C0} + 0.10h)^2$$

The acoustic intensity at the vibrating surface can be determined from: (acoustic output power of transducer) (surface area of transducer).

In application, however, because the threshold for cavitation is affected by the characteristics of the transducer (signal frequency, acoustic pulse length, acoustic uniformity / non-uniformity of the vibrating face) and by various conditions (depth of submersion, temperature, dissolved air content of the liquid), cavitation may not be initiated until the acoustic intensity is significantly higher, e.g., between 0.3 $(P_{C0} + 0.10h)^2$ and 0.4 $(P_{C0} + 0.10h)^2$.

Figure 5.22 Amplitude transformers (horns)

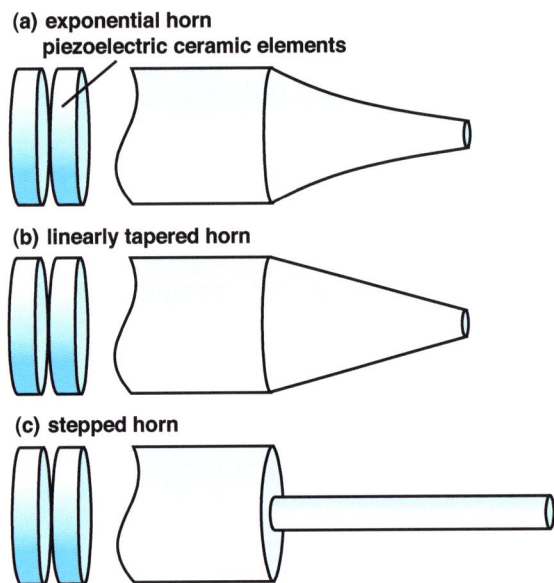

(a) exponential horn
   piezoelectric ceramic elements

(b) linearly tapered horn

(c) stepped horn

Additional Applications for Ultrasonic Vibrations

In addition to powering ultrasonic water baths, ultrasonic transducers are used for welding plastics, atomizing liquids, and drilling or milling ceramics, metals, or other materials.

In order to perform effectively certain tasks, including melting and welding plastics, the longitudinal vibrations on the end faces of a transducer must be amplified and concentrated onto a small surface area. A device used for this purpose is variously called an amplitude transformer, a sonotrode, or, simply, a horn (Figure 5.22).

Typically made of titanium alloy or aluminum alloy, and bolted to the end of the ultrasonic transducer, the horn must match the frequency of the transducer. Its length should be one-half the wavelength of the ultrasonic signal in the transducer at the resonance frequency, or multiples of the half-wavelength value.

Figure 5.23 Ultrasonic welding

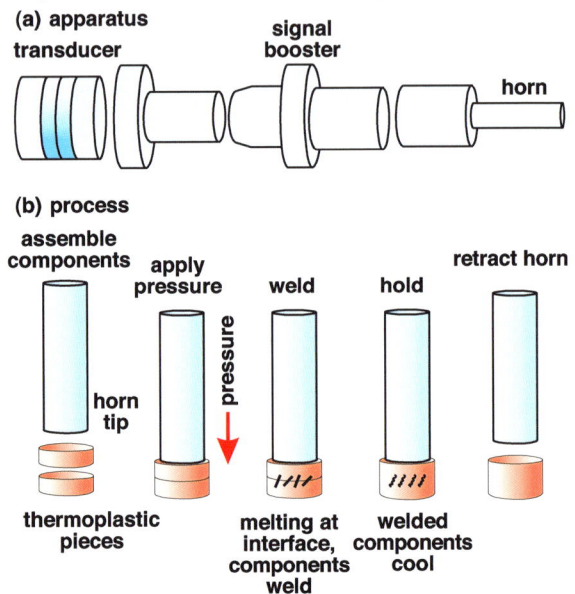

(a) apparatus

(b) process

Note that wavelength = sound velocity / frequency. Sound velocity in a piezoelectric ceramic element is approximately twice the frequency constant for the controlling dimension of the element, and frequency constants usually are among the product characteristics listed by ceramics manufacturers (e.g., see Table 1.6).

In practice, however, the propagation speed for the signal increases as the signal approaches the tip of the horn (because the surface area of the horn decreases), and this will make a horn constructed based on the simple wavelength—velocity—frequency relationship somewhat

Figure 5.24 Alternative approaches for atomizing liquids

(a) stepped horn (composite transducer)

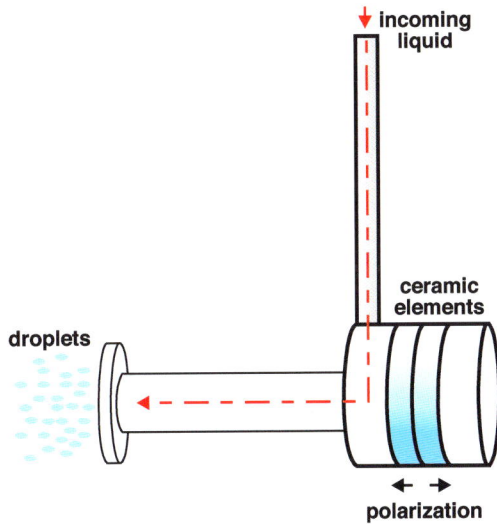

(b) submerged ceramic disk

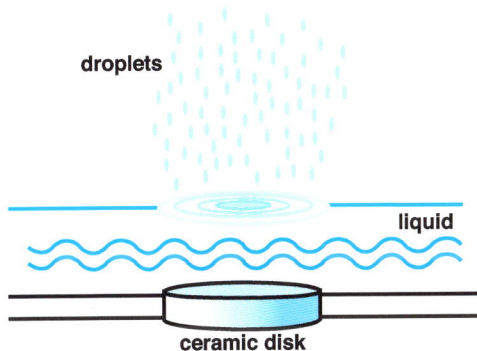

Ultrasonic vibrations in a piezoelectric ceramic element can be used to generate heat to melt and weld thermoplastics. In ultrasonic welding, a horn is brought into contact with one of the thermoplastic pieces to be joined, pressure is applied, and ultrasonic vibrations are passed through the first thermoplastic piece to the interface with the second thermoplastic piece. Frictional heat at the interface melts the material at the interface. When the vibrations are stopped, the temperature drops, the plastic solidifies, and the two pieces are bonded together (Figure 5.23). The process is fast and efficient, produces a strong bond, and eliminates the need for solvents and adhesives. Because the heat generated in this manner is confined to the area to be bonded, heat is rapidly dissipated, relative to alternative thermal-based approaches to bonding. Similar approaches are used to weld metals and bond wire mesh.

Ultrasound will effectively atomize liquids, and a stepped horn is well suited for this application. The liquid to be dispersed is pumped into the horn, and ultrasound from the transducer radiates into the liquid. Droplets separate from the surface of the liquid and exit from the tip of the horn (Figure 5.24a). Droplet size is a function of the frequency of the sound input—the droplets become smaller as the frequency increases:

longer than ideal. Energy also has been successfully transferred by connecting several one-half wavelength horns in series. In any case, the tip of the horn should vibrate with the maximum amplitude attainable for the configuration of the horn. Resonance frequency is little affected by the radial dimensions of a horn, provided these dimensions are less than one-fourth the wavelength of the ultrasound output of the transducer.

For an exponential horn, the vibrational amplitude increases by the same factor as the decrease in diameter. A tapered horn or a stepped horn usually is less effective at transferring energy, compared to an exponential horn, but these designs are viable alternatives because they are easier to manufacture.

Figure 5.25 Circuit for driving a submerged ceramic disk

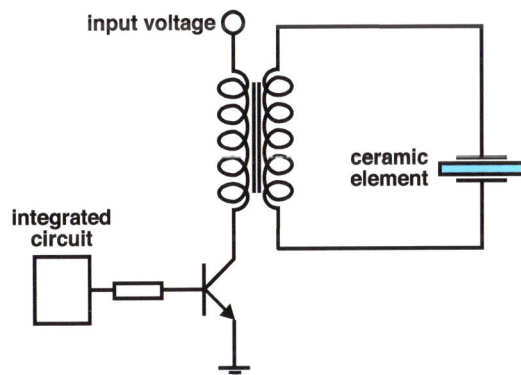

Figure 5.26 A piezoelectric ceramic transducer can act as a signal transmitter or a signal receiver

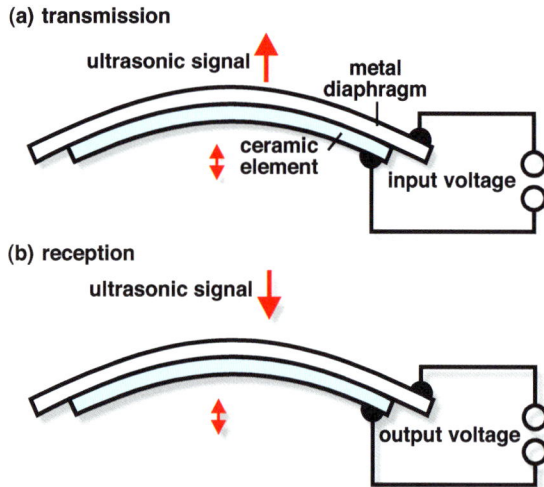

(a) transmission

ultrasonic signal

metal diaphragm

ceramic element

input voltage

(b) reception

ultrasonic signal

output voltage

Equation 5.12

$$\phi_p = 0.73 \sqrt[3]{(T_{surf} / (\rho f^2))}$$

where

ϕ_p = mean diameter of liquid droplets
T_{surf} = surface tension of liquid
ρ = density of liquid
f = frequency of input signal

In an atomization apparatus, liquid throughput increases linearly with increases in the surface area from which the liquid is radiating. A flare on the outlet end of the stepped horn effectively increases the surface area. Throughput also is a function of the amplitude of the ultrasonic vibrations, which in turn are a function of the voltage applied to the transducer. Increasing the size of the ceramic element will reduce the operating frequency, thereby increasing throughput, but at the cost of increasing the size of the droplets (Equation 5.12). Other factors limit throughput: an excessive temperature rise as a result of power dissipation in the transducer will put the ceramic element at risk for depolarization.

A simple, effective alternative approach for atomizing a liquid is to immerse a single ceramic disk of suitable characteristics, and approximately 1 mm thick, several centimeters below the surface of the liquid (Figure 5.24b). Application of a voltage causes a well-chosen disk to vibrate at megahertz

frequencies, and compression waves generated by this extremely high rate of vibration expel droplets from the surface of the liquid. Again, droplet size is governed by the signal frequency (Equation 5.12). A signal frequency of approximately 2 mHz will create water droplets sufficiently small to be airborne, and transducers operating at this frequency are used in air humidifiers. Figure 5.25 shows a simple circuit used to drive a ceramic disk for atomizing a liquid.

Transmitting Ultrasonic Signals

Because a piezoelectric element can convert electrical energy into mechanical energy or mechanical energy into electrical energy, an ultrasonic transducer is capable of both generating and receiving a signal (Figure 5.26). This principle has been adapted to measuring distances in air, water, or other fluid media, determining flow rates, and other applications. Although a single ultrasonic transducer can both generate and receive a signal, the two functions often are separated to optimize the performance of each task. Thus, a distance measuring system typically will include a transmitting transducer and a receiving transducer, but there are exceptions to this generalization.

Directivity

One of the important parameters of a signal-transmitting / signal-receiving transducer is its directivity, or beam pattern, which is the response level of the transducer as a function of the in-

Figure 5.27 Short excitation pulses ensure accurate detection at short distances

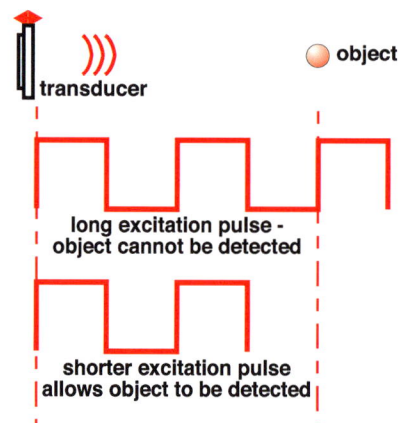

object

transducer

long excitation pulse - object cannot be detected

shorter excitation pulse allows object to be detected

cident angle of the sound. The directivity of a transducer, a function of the aperture of the transducer and the frequency of the signal, will be the same whether the transducer is operating as signal transmitter or signal receiver. Directivity is used to calculate the power and efficiency of the transducer. A high directivity indicates that as a transmitter the acoustic power of the transducer will be concentrated into a small region, and as a receiver it will allow high image resolution and reduce the influence of noise.

The directivity index, DI, quantifies the sharpness (directionality) of the signal pattern and is closely associated with the beamwidth, BW. Approximations for the half-power (-3 dB) directivity index and beamwidth for a transducer whose dimensions are greater than one wavelength (λ) of sound in water, are:

Equation 5.13

$$DI = 10 \log (4\pi)(\text{area} / \lambda^2)$$

or

Equation 5.14

$$DI = 10 \log (2l / \lambda)$$

and

Equation 5.15

$$BW \cong 60\lambda / d$$

or

Equation 5.16

$$BW \cong 50\lambda / l$$

where

area = surface area of a circular or square-faced element

l = length of a thin cylinder or length of a side of a square-faced element

Beamwidth for the -10 dB level and between the first nulls are approximately 1.8 times the -3 dB value and approximately 2.3 times the -3 dB value, respectively.

Measuring Distances in Air

Devices in which piezoelectric transducers are used to measure distances in air, collectively called impulse-echo applications, include intruder

Figure 5.28 Circuits for operating a transducer as a receiver or transmitter

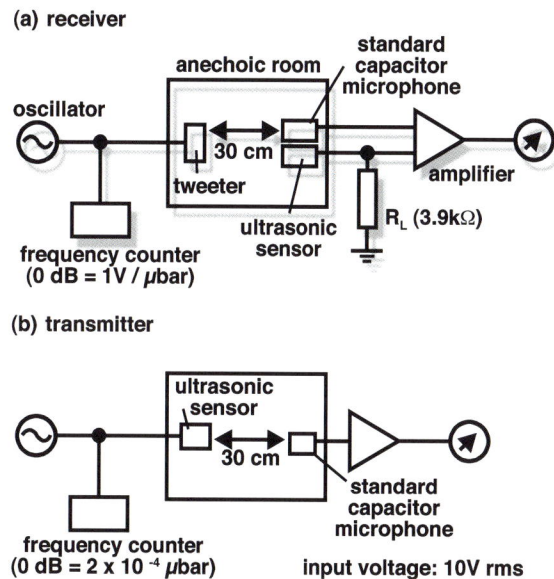

(a) receiver

frequency counter
(0 dB = 1V / μbar)

(b) transmitter

frequency counter
(0 dB = 2 x 10^{-4} μbar)

input voltage: 10V rms

alarms, fill-level monitors for grain silos or other large containers, and proximity-warning devices (parking aids) that guide a driver's attempts to park a truck or automobile.

Any impulse-echo application requires compromises based on emphasizing the most important factors for the specific application. The operating range for an impulse-echo device is a function of two factors: the operating frequency of the transducer and the power the transducer generates. The latter factor, power output, depends on the mechanical characteristics and thermal constraints of the device. The operating frequency of the transducer requires careful consideration because reflection and absorption of ultrasonic signals are frequency dependent. For long range applications, low frequencies offer an advantage, because signal damping is much less. (Signal damping increases significantly as the signal frequency increases: in air, sound pressure for a 20 kHz signal drops to half its original value over a distance of 10 m; sound pressure for a 50 kHz signal drops to half its original value over 3 m.) For good signal directivity and effective resolution of an object, however, the frequency should be as high as possible. Also, detection will be unreliable if the reflecting surface of the object to be detected has the same dimension as the wavelength used for the signal. In this situation, the signal frequency must be increased. For accurate measurement of distances, especially at short ranges, the excitation pulses must be short (Figure 5.27).

If necessary, the range of an impulse-echo device can be extended by narrowing the bandwidth, or by using reflectors to focus the signal. A narrow bandwidth offers an additional benefit: the input power requirement is lower than the power requirements for wide-band transducers. On the other hand, a broader bandwidth will exhibit less noise.

Air temperature, humidity, and cleanliness affect the practical range of signal transmission by an ultrasonic transducer. Air currents can produce sudden changes in the character of the signal path (temperature, particle content, etc.), and therefore also are an important factor affecting the signal range.

Flexional Ultrasonic Transducers for Determining Distances in Air

Just as a piezoelectric ceramic disk bonded to a thin, flexible metal diaphragm can transmit audible sound, such a mechanism can be designed to transmit or receive ultrasonic signals. Using one of the designs in Figure 5.29, and matching the applied voltage to the mechanical resonance frequency of the flexing element, will maximize the deflection of the flexing element and thereby will maximize the ultrasound signal. The mechanical resonance frequency is a function of the characteristics of the piezoelectric ceramic disk, the thickness and diameter of the metal diaphragm, the manner in which the flexing

element is mounted, and the manner in which the transducer is incited to vibrate. For reliability, values for mechanical resonance frequency should be determined by measurement. Because mechanical resonance frequency depends on the thickness and diameter of the metal diaphragm, it can be changed for the device in Figure 5.29c or Figure 5.29d simply by changing one or both of these dimensions of the diaphragm. The dimensional relationship between the ceramic disk and the metal diaphragm (Figure 5.29) must be maintained, however.

The device in Figure 5.29a must be mounted in a manner that prevents either the outer area (beyond the nodal ring) or the central area of the flexing element from radiating sound. If the entire surface is allowed to radiate, counterphase signal produced at the outer circumference will partially suppress the signal from the central area. The device in Figure 5.29c can be used at frequencies as high as 40 kHz, or higher, and, because the wavelength of the ultrasound signal in air is small, relative to the diameter of the membrane (see Equation 5.17), this design offers sharp directivity and good beam characteristics.

The open designs of the devices in Figure 5.29a and Figure 5.29b limit the use of these devices to relatively clean and dry atmospheres, but the devices in Figure 5.29c and Figure 5.29d can be enclosed, to adapt them to applications in dusty or humid situations.

A narrow ultrasound beam minimizes interference and reflections from objects outside the beam path. If the beam is too narrow, however, air conditions can divert the beam from its

Figure 5.29 Flexional ultrasonic transducers

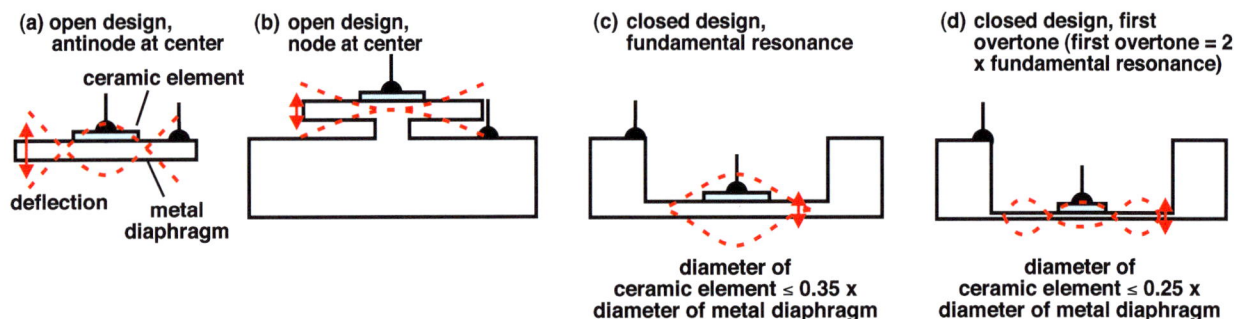

(a) open design, antinode at center
ceramic element
deflection
metal diaphragm

(b) open design, node at center

(c) closed design, fundamental resonance
diameter of ceramic element ≤ 0.35 x diameter of metal diaphragm

(d) closed design, first overtone (first overtone = 2 x fundamental resonance)
diameter of ceramic element ≤ 0.25 x diameter of metal diaphragm

Figure 5.30
Directivity of a flexional
ultrasonic transducer

(distance to receiver > $D_{\varrho}^2 / 2\lambda$)

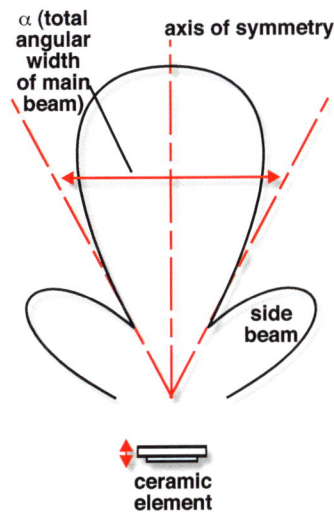

α (total angular width of main beam)

axis of symmetry

side beam

ceramic element

Equation 5.17

$$\sin \alpha / 2 \cong \lambda / D_{\varrho}$$

Equation 5.18

$$\sin \alpha / 2 \cong v_{air} / D_{\varrho}f$$

where

α = angular width of main
     ultrasound beam

λ = wavelength of ultrasound signal in air

D_{ϱ} = diameter of area radiating ultrasound

v_{air} = velocity of sound in air
     (~344 m/s)

f = frequency of input signal

    As λ is decreased, or D_{ϱ} is increased, the angle of the beam narrows. Directivity is sharpened and the range is lengthened. On the other hand, if λ is increased or D_{ϱ} is decreased, to the extent that λ equals or exceeds D_{ϱ}, the directional characteristic assumes a spherical form, to which equations 5.17 and 5.18 do not apply.

Transmitting Ultrasonic Signals in Water: Echo Sounders and Related Apparatus

Transmitting and receiving acoustic signals in water—echo sounding—is significantly differ-

intended path, particularly at longer distances between the source of the beam and the receiver. Directivity of a flexional ultrasonic transducer depends on the wavelength, λ, of the emitted ultrasound in air, the diameter, D_{ϱ}, of the area that is radiating the signal, and the uniformity of the vibrations across the surface of the ceramic element (there should be a narrow distribution for vibrational amplitude). When λ < D_{ϱ}, the angular width of the main ultrasound beam (Figure 5.30) can be determined from these relationships:

ent from the corresponding activities in air and, consequently, a transducer intended for an underwater application is designed differently and is constructed of different materials. The type of construction of a device depends on the signal frequency required, and signal frequency depends on the intended application. Signals from a few hertz to approximately 1 kHz are used for ocean acoustics, signals between approximately 1 kHz and 100 kHz are used for strategic (weapon / detection) sonar, signals between approximately 50 kHz and 1 MHz are used in fish-finding and mine-detecting devices, and for bottom profiling, and signals from approximately 1 MHz to values in the gigahertz range are used for medical ultrasound imaging and nondestructive structural testing, and in ultrasonic microscopes. Ultrasonic transducers that operate at frequencies greater than 20 MHz provide the highest resolution images, and thus improve diagnostic accuracy; design of transducers in the 30-100 MHz range is a current challenge.

    The first practical application of piezoelectric principles and materials was an echo sounding device for detecting submarines, developed just after World War I. Typically, in either an in-liquid or an in-air echo sounding application a single, electronically switched piezoelectric transducer provides both the transmitting function and the receiving function (Figure 5.31). When the transducer is in the transmitting mode, an input voltage flexes the piezoelectric element and causes it to emit ultrasonic waves. When the

Figure 5.31 Echo sounding system: piezoelectric ceramic transducer alternates between signal transmitter and receiver

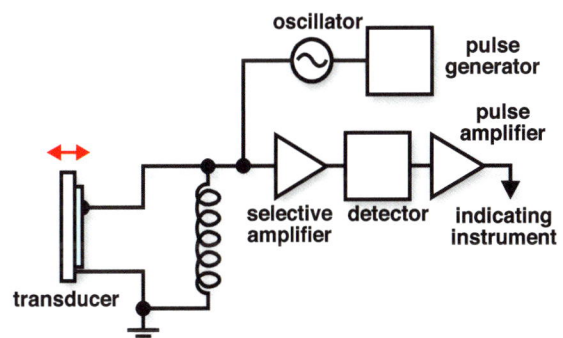

oscillator

pulse generator

pulse amplifier

selective amplifier

detector

indicating instrument

transducer

transducer is switched to the receiving mode, the returning ultrasonic signal flexes the piezo-electric element and the element produces an output voltage. To locate an object in air or water, the transducer is used to transmit a short pulse of ultrasound in the general direction of the object. The signal is reflected by the object and is picked up by the transducer, now acting in its receiving role (Figure 5.32). The distance from the echo sounder to the object is determined from the speed of the transmitted waves in the medium and the angle formed by the transmitted and reflected signal:

Equation 5.19

$$L = ((v)\,(t)\,(\cos \beta)) / 2$$

where

L = distance to object

v = velocity of transmitted signal

t = time between transmission and return of signal

2β = angle formed by transmitted and returned signal

If the transmitter and receiver functions are accomplished by the same transducer, or if separate transmitter and receiver transducers are used, but they are in close proximity, cos β approximates 1, and distance depends only on the velocity of the signal and the time between its transmission and its return.

Figure 5.32 Locating an object with ultrasound

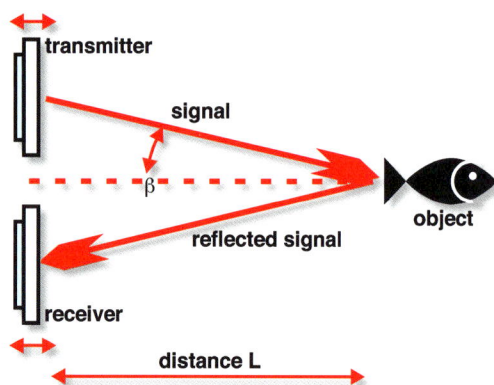

Simple and reliable, echo sounding devices are widely used in nautical applications, for measuring water depth, locating underwater hazards, or detecting fish. Equivalent in-air applications include intruder alarms and the proximity-warning devices intended to reduce the guesswork involved in parking an automobile (either a single transducer is charged with both the transmitting and receiving roles, or the roles are assigned to separate transducers). A proximity sensor for in-air applications is created by setting a detection threshold in the receiver circuitry of the echo sounder and using the object to be detected as a reflector (Figure 5.33). In security applications, the transmitted signal is directed at a mirror, and an object interfering with the beam will be detected.

In any in-liquid or in-air echo sounding application, vibrations established during signal transmission must subside before the transducer can differentiate a returning signal from background, so effective operation dictates that there be at least a minimum distance between the transducer and the object. For example, an echo sounder for in-air applications, operating at a frequency of 200 kHz, will be unable to discriminate between an object and noise if the object is within a distance of approximately 0.2 m of the signal source. Furthermore, the excitation impulse of the ultrasonic signal should be as short as practical (Figure 5.27), and this dictates the use of a wide bandwidth and a high operating frequency. Echo sounding devices for medical applications are operated at megahertz frequencies, for example, to ensure good definition of objects only a few millimeters away. A high operating frequency also means a smaller transducer can be used, so the echo sounder can be made more compact. On the other hand, for a given distance, signal damping increases significantly as the signal frequency increases.

In nautical applications of echo sounders, at distances of less than 10 m, echo intensity depends almost exclusively on distance and the reflection coefficient of the target object. For distances greater than 10 m, signal absorption by the water, in both the transmission and return

Figure 5.33 Piezoelectric transducer used as a proximity sensor

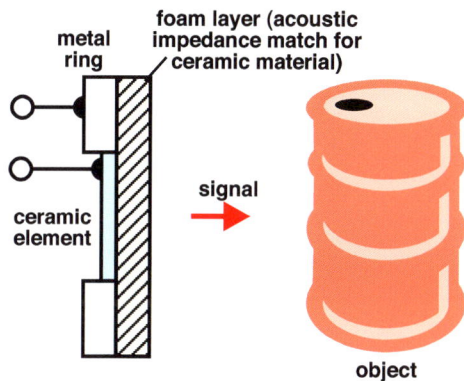

directions, also must be considered. For a 200 kHz signal, absorption is approximately 0.06 dB/m. Consequently, if an object is 100 m from the transducer (200 m total path length), absorption totals 0.06 x 200, or 12 dB.

The maximum range for an echo sounder depends on the construction of the transducer (e.g., Figure 5.33 or Figure 5.34). At an operating frequency of 200 kHz a typical maximum range for in-water applications is approximately 100 m. At the megahertz frequencies employed in medical applications, signal attenuation limits the maximum range to a few centimeters, or less, but longer maximum range is irrelevant to these applications.

In an echo sounder used as a proximity sensor (Figure 5.33) the ceramic disk is surrounded by a metal ring (e.g., aluminum alloy) and the signal-transmitting surface is covered with a layer of foam of matching acoustical impedance. Ultrasound radiates only from the surface contacting the foam, ensuring high directivity and a narrow beam width, α.

Similarly, to ensure good performance from an echo sounder in nautical applications, the transducer should be housed as shown in Figure 5.34. The back surface of the ceramic disk should be isolated from contact with the water, but for greatest mechanical strength, and to minimize radial resonances that could interfere with the operating resonance, the periphery of the disk cannot be isolated. For optimum performance, the interface protecting the signal-transmitting

face of the transducer should have an acoustic impedance intermediate between that of the ceramic and that of water. The bandwidth for the signal can be significantly increased by acoustically matching the thickness of the interface to the wavelength of the signal (the thickness of the interface layer should be 1/4 the wavelength of the transmitted signal in the interface material). Foam rubber is a good choice for isolating the back surface of the ceramic disk because it is an acoustical equivalent to a layer of air. Many epoxy resins, plastics, or other materials meet the criteria for an ideal interface. Of course, the molding compound used to suspend the components of the echo sounder in place should exhibit minimal sound absorption.

In nautical applications, as well as in air applications, the range of an echo sounder can be increased, and small objects can be more easily detected, by narrowing the signal beam. As the beam is narrowed, however, the intensity of the reflected signal will fall off more sharply, and the perceived distance will increase, whenever the motion of the boat establishes an oblique angle between the transducer and the target object, whether the sea bed, a submerged object, or a fish. Note that the effective beam width of the transducer is affected by operating the transducer in both the transmitting and the receiving functions: a 6 dB signal transmitted from the transducer will broaden to 12 dB as it returns. The

Figure 5.34 Piezoelectric echo sounder in a well designed housing

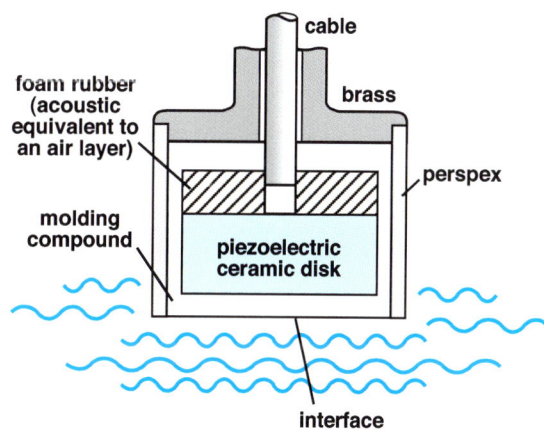

directivity characteristic should have the smallest possible side beams (Figure 5.30); large side beams can make the main beam ineffective. Also, the signal must radiate downward only, or the boat's wake, other surface turbulence, or even the boat itself could create interference.

Important characteristics of a piezoelectric transducer for echo sounding include the diameter of the piezoelectric ceramic element (a larger diameter element produces a narrower beamwidth), its operating frequency (which usually is the series resonance frequency), its impedance at the operating frequency, the minimum pulse duration that it can transmit (which depends on the bandwidth), the directivity of its signal, and its sensitivity (determined by measuring the response of the transducer to a totally reflected signal at a known distance). Parallel inductance tuning ensures a good match between the transducer and the load, and improves the bandwidth.

The performance characteristics of the ceramic element are key to the design of an echo sounding device. At first glance either a tall, narrow-diameter cylinder, vibrating in the axial mode (d_{33} mode), or a short, broad disk with a large diameter:thickness ratio, vibrating at the thickness resonance, apparently could supply the 150-200 kHz vibrations needed for short range (< 10 m) or medium range echo sounding, but there are complications. For a tall cylinder, the acoustic wavelength would be large, relative to the radiating area ($\lambda > D$, see Equation 5.17), and this would produce an almost spherical directivity characteristic. Alternatively, the diameter of a short, broad disk would have to be inconveniently large to provide a true thickness resonance at a suitable frequency.

By elimination, the drawbacks to using a tall ceramic cylinder or a short, broad disk for echo sounding applications dictate that the ceramic element be a disk whose thickness and diameter are comparable. The output from such an element will be more complex, relative to the output from a tall cylinder vibrating in the axial mode or a broad disk vibrating at thickness resonance. There will be coupling between vibration in the radial direction and vibration in the thickness direction, and there will be several resonances within the required frequency range, but one of these resonances will produce a suitably directive beam. The frequency of this resonance will be dictated largely by the thickness of the disk, but the directivity will be dictated largely by the diameter of the disk, and also by the operating frequency and by the variation of vibration amplitude across the disk surface. A thickness:diameter ratio of approximately 0.4 will produce an operating resonance with the best separation from the other resonances; a simple, compact transducer can be constructed from such a disk.

Flexional ultrasonic transducers

Low Frequency Transducers

Several types of transducers are suitable for providing low frequency signals (less than 100 Hz to approximately 3-5 kHz) for underwater applications, including flexional disks, flextensional transducers ("moonies" and "cymbals"), and Helmholtz resonators (2). The basic requirement of a transducer producing a low frequency signal is that it efficiently produce large displacements. The amount of sound generated is small, because the dimensions of these transducers are small, relative to the wavelength of the sound they generate (at a frequency of 1 kHz, sound waves in water are approximately 1.5 m long; at lower frequencies, the wavelength is considerably longer), but this is acceptable because the power requirements at these frequencies also are small. An effective low frequency transducer will be more compliant than a ceramic disk or stack vibrating in the thickness mode, and to attain this compliance, a low frequency transducer can incorporate flexional movement.

Flexional transducers can be designed to transmit or receive low frequency ultrasonic signals in an underwater environment, as effectively as in air. In a bilaminar transducer the ceramic disk that is polarized in the same direction as the electrical field expands and the disk that is polarized in the direction opposite the electrical field contracts, causing the element to bend. This converts the small displacement in each ceramic disk into a larger displacement of the surface that is radiating ultrasound into the water. Variations on the basic construction include inserting a flexible metal disk between the two ceramic disks, or replacing one of the ceramic disks with a metal disk. Relative to a single bilaminar disk, a double bilaminar disk (two bilaminar disks radiating in opposite directions, and separated by a spacer ring) will increase efficiency. The simple construction of these devices makes them economical, and they exhibit high sensitivity in non-demanding applications.

A flextensional transducer converts the small displacements of thickness mode vibration in the ceramic element into larger displacement by the concave metal, plastic, or glass-reinforced plastic shell that surrounds the ceramic element (Figure 5.35). The physical dimensions of a flextensional transducer usually are very small, relative to a wavelength of sound in water at the resonance frequency of the transducer (typically 300 Hz to 3 kHz). Because of this, a flextensional transducer will produce an omnidirectional signal, in the plane perpendicular to the axis of the ceramic element. Flextensional transducers generally offer a higher power output than flexional disks and, compared to a flexional disk of comparable dimensions, the mechanical quality factor for a flextensional transducer can be two to four times lower (values can be as small as $Q_m = 2$). Lower Q_m values indicate flextensional transducers have

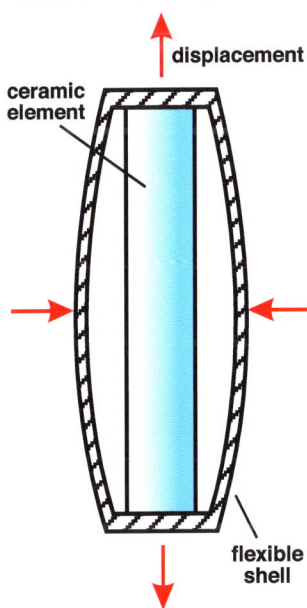

Figure 5.35
Flextensional transducer

displacement

ceramic element

flexible shell

superior signal-transmitting / signal-receiving capability.

Although the performance of flextensional transducers generally is superior to that of flexional transducers, the design of a flextensional transducer requires meticulous attention, to properly balance all of the interactions among the various contributing parameters. For example, the resonance frequency of the transducer and the hydrostatic pressure the transducer can function under depend on the material used to construct the shell, the thickness of the shell wall, the ratio of the dimensions of the major axis and the minor axis, and, to a lesser extent, the nature of the ceramic material and the construction of the ceramic element. Further, to ensure a usefully high output, the ceramic element must be under a compressive pre-stress. Pre-stress usually is introduced by squeezing the shell at the minor axis, to elongate the shell along the major axis, then inserting the ceramic element into the shell. Relaxation of the pressure on the shell produces pre-stress on the element. The pre-stress must account for partial compensation by hydrostatic pressure under operating conditions.

A Helmholtz resonator with a flexional disk at the rear surface of the cavity can be a compact and effective low frequency signal transmitter, and, if constructed properly, can function in this capacity at considerable depths. The resonance frequency for a Helmholtz resonator in water is approximately four times its resonance frequency in air, and the mechanical quality factor, Q_m, usually exceeds 20 (2).

Higher Frequency Transducers: Tonpilz Transducers

Tonpilz transducers are metal end / ceramic center / metal end composite, longitudinally vibrating transducers designed for generating signals over a broad frequency range—from approximately 3 kHz to approximately 100 kHz. When incorporated into arrays, they form signal patterns that can be adjusted to various angles for locating and imaging objects. The name is a combination of the German words for "sound" and "mushroom", and is well suited to describing the function and general shape of these devices

(Figure 5.36). Although the specific construction can vary (for sources of descriptions, see reference 2), a well-designed tonpilz transducer will achieve high radiated sound power and can have an efficiency of 80-90%. A tonpilz transducer is operated near its resonance frequency.

The radiating face of the head mass interfaces with the water environment. The surface area of the head mass typically is larger than the surface area of the ceramic element(s) because this increases the power output for a given strain in the ceramic and ensures a better impedance match between the active transducer element(s) and the water environment. However, the ratio for the surface area of the head mass to the surface area of the ceramic element(s) is limited to a range of 3-10:1 by the need to avoid flexional resonances in the head mass, and the need for a sufficient volume of ceramic for activation. If the diameter of the face of the head mass is equal to one-half the wavelength of the signal in water, multiple-unit arrays can be operated in phase, and the signal beam can be directed. The weight of the head mass is minimized by using aluminum, magnesium, or other low density metal as the construction material, and minimal thickness is determined by setting the flexional resonance at about twice the fundamental resonance. The threshold for cavitation on the face of a tonpilz

Figure 5.36 Tonpilz transducer

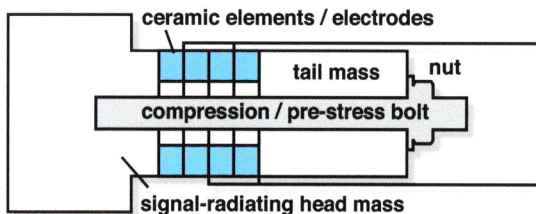

transducer can be determined from Equation 5.9 or 5.10.

Important factors affecting the performance of the ceramic stack include the number of elements, the dimensions of the elements, the compliance, electrical field limits, power-handling capability (e.g., level of dielectric and

mechanical losses, ability to dissipate generated heat), electrical impedance, and electromechanical coupling factor of the ceramic, and cost. If the transducer is intended for both transmission and reception, a Navy type I ceramic (hard lead zirconate titanate, Curie point ≥310°C) is suitable; if a high acoustic field is required, a Navy type III ceramic (very hard lead zirconate titanate, Curie point ≥290°C) is needed to minimize the heating effects of high power input.

For low impedance, high power applications, in which the transducer must both transmit and receive, the ceramic stack typically consists of 4 or 6 elements. The paired ceramic elements can be connected in parallel, with alternating polarity and electrical fields. The maximum voltage that can be applied to the transducer is: (electrical field limit) (thickness of one ceramic element).

The tail should have as light an acoustic load as possible (e.g., air) and, to maximize the displacement of the head, the tail mass should be as large as practical, and 3-5 times the mass of the head. Commonly a high density metal such as tungsten, stainless steel, or brass is used.

The head, center, and tail portions of a tonpilz transducer are bonded together and compressed by a central bolt which places a compression pre-stress of 14-41 MPa on the ceramic elements, so that high drive voltages do not put the ceramic elements under tension.

High Frequency Transducers
Although a longitudinally vibrating tonpilz transducer can be constructed to generate signals at frequencies higher than 100 kHz, the dimensions for a tonpilz device are impractically small at these frequencies. It is more convenient and more economical to construct a transducer from a single, thin ceramic element, with both polarization and the electrical field in the thickness dimension. Such an element will vibrate in the thickness mode, but also may exhibit unwanted vibrations within the ceramic element itself. Alternatively, signal frequencies higher than 100 kHz can be generated by a 1-3 composite of lead zirconate titanate in an inactive support, or by a

sheet of polyvinylidene difluoride, with little or no vibration in the signal-generating structure (see **Composites**; also see **PVDF** in the **Introduction**).

Ultrasound Transducers for Medical Diagnostics

An obstacle to developing medical ultrasound transducers is the large acoustic impedance mismatch between the piezoelectric ceramic signal-producing element and the human tissue or water load through which the signal is transmitted. If this mismatch is not compensated for, most of the acoustical energy generated by the ceramic element will be reflected back and forth between the rear face and the front face of the element. In this unmodified system the pulse transmitted into the tissue will be long, and axial resolution will be poor.

A common response to this problem, based on traditional nautical echo sounding apparatus (Figure 5.34), has been to apply an acoustically matched backing material to the rear face of the ceramic. This approach shortens the pulse into the tissue and thus improves resolution, but sacrifices sensitivity. A better solution to the acoustic impedance mismatch is to apply a minimal backing to the ceramic element and incorporate an acoustically matching layer at the front face of the element. This combination ensures both a suitably short pulse and useable sensitivity. More than one matching layer can be applied to the front face, and the layers can exhibit various combinations of acoustic impedance and thickness (3). A lens in front of the entire assembly will focus the ultrasonic signal at a specified distance.

Composite materials also are being used to overcome the acoustic impedance matching problem between signal-generating piezoelectric ceramic elements and water or other media (see **Composites**).

Radially or tangentially polarized miniature (<5 mm) hollow, lead zirconate titanate spheres meet the size (smaller than the acoustical wavelength of interest), directionality (they produce an omnidirectional signal), acoustic impedance, and other requirements for hydrophonic and medical ultrasound applications (4). The piezoelectric charge constant for these miniature spheres, 600 to 1800 pC/N, is substantially higher than for bulk element lead zirconate titanate transducers. This is an important value, because the signal to noise ratio for a hydrophone can be gauged from the product: (hydrostatic piezoelectric charge constant) x (hydrostatic piezoelectric voltage constant), and noise is a limiting factor in hydrophone design. The miniature piezoelectric spheres can be produced, inexpensively, in large numbers.

PZT / Relaxor Material for Medical Diagnostics Applications

Piezoelectric, chemical, and other characteristics of traditional PZT-type materials developed for use in accelerometers, underwater transducers, or other applications often do not meet the demanding requirements for medical diagnostic transducers: very high permittivity, sensitivity, and electromechanical coupling factors, and a high Curie point. Lanthanum-containing modifications of these materials (PLZT) have the necessary characteristics, but are difficult to process consistently without sintering additives or complicated post-sintering treatments that increase the price of the product and reduce the reproducibility of its performance and aging properties. Further, these materials are susceptible to corrosion in environments encountered in medical diagnostics.

Combinations of a conventional PZT material and PNN, a relaxor material comprised of lead, nickel, and niobium ($Pb(Ni_{1/3}Nb_{2/3})O_3$), produce very soft piezoceramics that are well suited for medical imaging and other medical diagnostics. Desirable characteristics of a PZT-PNN material include very high permittivity and coupling factors, e.g.:

- relative permittivity ≥3800
- planar coupling coefficient ≥60%
- thickness coupling coefficient ≥45%
- Curie point ≥200°C
- dense structure with fine-grained, uni-modal crystals (Figure 5.37), suitable for dicing into fine sub-elements for arrays
- high corrosion resistance in water-based and organic solvents

These characteristics are comparable to those for APC 850, APC 855, and APC 856 PZT materials (Table 1.6). A PZT-PNN material also is suitable for ink-jet applications and other applications in which high sensitivity, low porosity, and a small grain size are required, and highly reproducible batch to batch performance is important.

A tape-cast disk of a PZT-PNN material with the following characteristics was used to develop a computer model of an imaging transducer:

thickness: 72 μm
density: 7420 kg/m^3
electromechanical
coupling factor (k$_t$): 43.1%
antiresonance frequency: 33.9 MHz
relative dielectric constant: 1920
mechanical losses: 4.7%

Figure 5.37
Microstructure of a PZT-PNN material

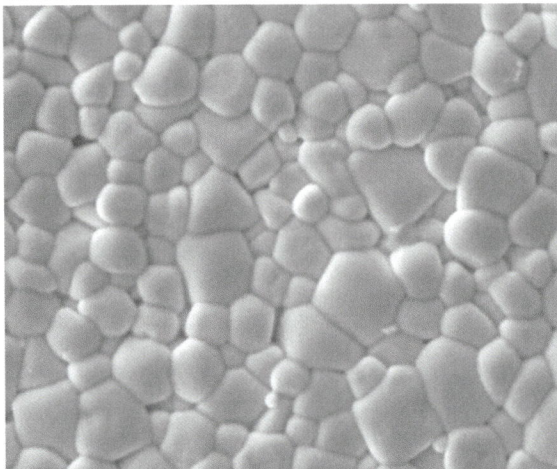

Original magnification: x5000
Mean grain size: approximately 2 μm

Properties of the simulated transducer include:
frequency at center: 31.9 MHz
active area: 0.5 mm^2
thickness of matching layer: 17.5 μm
acoustical impedance
of matching layer: 4.2 MRa*
acoustical impedance
of backing: 7.7 MRa
serial inductance: 19 nH
* 1 megarayleigh = 10^6 kg / (m^{2}s)

Figure 5.38 shows the impulse response profile for the simulated transducer. After only approximately 0.15 μs, the amplitude of the signal has fallen 20 dB. From the velocity of sound in water, 1483 m/s, this pulse duration indicates the wave would have traveled a pathlength of ~216 μm—a measure of the spatial resolution obtainable with the transducer. Performances are highest when the active area is less than 3 mm^2. The acoustical impedance of the backing does not appear to have great influence on the characteristics of the transducer, and variations here can be compensated for through adjustments to the matching layer. Figure 5.39 is the frequency response plot for the transducer. The -6 dB bandwidth is 39%.

Flow Meters

Two piezoelectric ceramic transducers, positioned a distance apart in a flow path, can be used to measure the velocity of the gas or liquid in the flow path (Figure 5.40). The two transducers simultaneously transmit short pulse signals, are switched to the receiving mode, receive the signal from the upstream / downstream counterpart, and are returned to the transmitting mode. If the medium in the flow path is static, the delay times between signal transmission and reception in the upstream direction and signal transmission and reception in the downstream direction are equal. A medium in motion, however, will accelerate the downstream pulses and delay the upstream pulses in proportion to the velocity of the flow. Acceleration / delay times can be determined from Equation 5.20; flow velocity can be determined from Equation 5.21:

Equation 5.20

$$t_1 = L / (v + w)$$

and

$$t_2 = L / (v - w)$$

Equation 5.21

$$w = (L / 2) (1 / t_1 - 1 / t_2)$$

where

t_1 , t_2 = delay times

L = measuring distance

v = sound velocity in medium

w = flow velocity

Figure 5.38 Impulse response profile for a simulated high frequency transducer incorporating a PZT-PNN ceramic disk

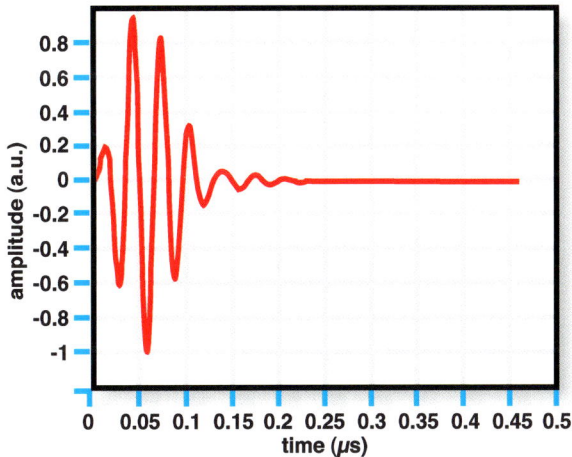

Note that by using equation 5.21, the flow velocity can be calculated without knowing the velocity of sound in the medium.

If the transducers are positioned as depicted in Figure 5.40, a value for flow velocity calculated from Equation 5.21 will be valid only for the gas or liquid flowing along the center of the pathway. Positioning the transducers as depicted in Figure 5.41 ensures the ultrasound signals will traverse the spectrum of flow velocities within the pathway, and acceleration and delay values calculated by using this configuration will be a measure of the mean flow velocity, w_{mean}, through the pathway. This, in turn, is a measure of the volume passing through the pathway.

Figure 5.39 Frequency response profile for a simulated high frequency transducer incorporating a PZT-PNN ceramic disk

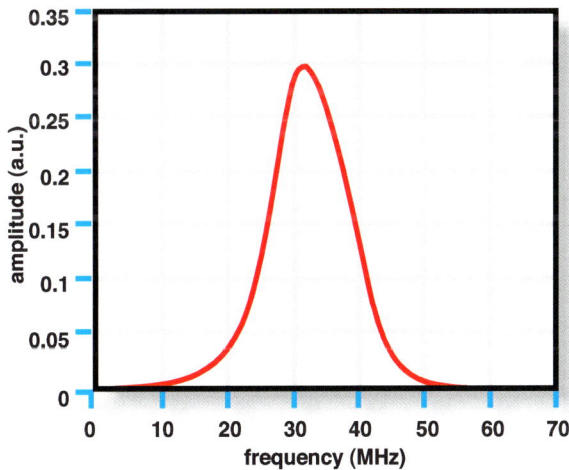

Figure 5.40 Ultrasonic flow meter

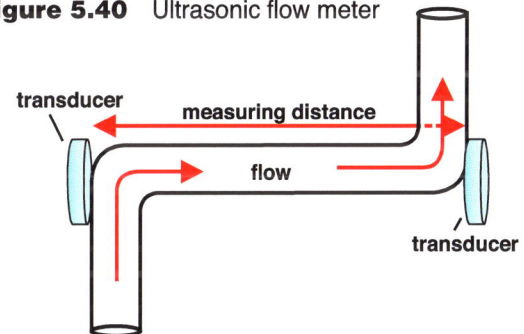

Figure 5.41 Ultrasonic flow meter configured to measure mean flow velocity

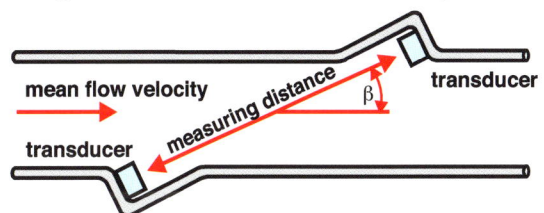

Equation 5.22

$$w_{mean} = L / 2 \cos \beta \, (1/ t_1 - 1/ t_2)$$

where

L = measuring distance
$\cos \beta$ = cosine of angle at which ultrasound
beam traverses flow pathway
t_1 , t_2 = delay times

In any medium, sound velocity varies with temperature, and temperature fluctuates over time. Equations 5.21 and 5.22 do not account for changes in temperature, and, therefore, apply only when measurements t_1 and t_2 are made simultaneously. If this condition cannot be met, other measuring principles must be used. For some situations, however, it might be useful to know that sound velocity in water can be made independent of temperature between 5°C and 45°C by mixing 18% ethyl alcohol by volume with the water (5).

Figure 5.42 depicts an alternative approach to having both transducers perform in the same mode simultaneously. In this approach one transducer acts as the transmitter while the other acts as the receiver, then the roles are reversed. Under control of the clock, the selector determines the role of each transducer.

Fluid Level Sensors

Fluid level sensors are used to monitor levels of liquids exhibiting a wide range of complexity and viscosity, from water to gasoline to oil or hydraulic fluids. These devices are keyed to the difference in

Figure 5.43 Fluid level sensor

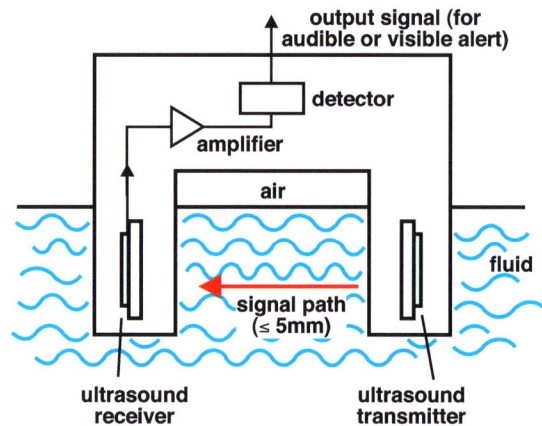

ultrasound transmission through the liquid being monitored versus ultrasound transmission through air (Figure 5.43). When there is liquid between the transmitting transducer and the receiving transducer, acoustic coupling is high; when the fluid level falls and air occupies the signal path, coupling is low, and an output signal is relayed to an audible alarm, a dashboard light, etc.

Delay Lines

In various applications electronic systems require delay of an electrical signal for as long as several milliseconds without sacrificing information in the signal. Electronic transmission lines cannot delay a signal without also attenuating and distorting it. In a piezoelectric acoustic delay system a transducer converts the electrical signal to an acoustic signal, passes the acoustic signal through

Figure 5.42 Ultrasonic flow meter with transducers acting in complementary roles

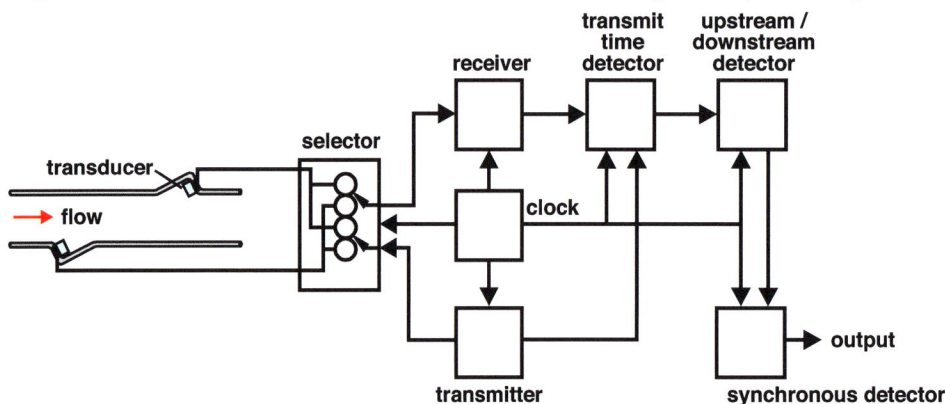

Figure 5.44 Piezoelectric delay line

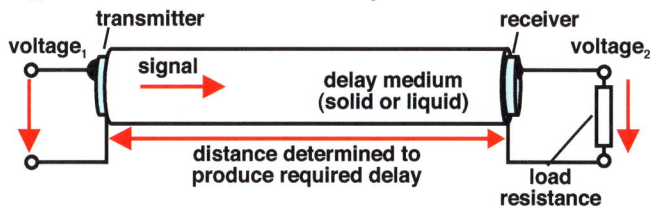

a delay medium, then converts the acoustic signal back to an electrical signal (Figure 5.44). At each step, energy conversion is highly efficient.

Key Finders

A key finder is a compact transmitting / receiving device designed to be attached to a key ring. Relative to the sequence of events in other applications, in a key finder the sequence of the transmitting and receiving functions is reversed. A key finder reacts to receiving an audible signal from its owner (approximately 1 kHz) by emitting a shrill 3 to 4 kHz response, enabling the owner to locate the key ring. Figure 5.45 shows the basic construction of a key finder is equivalent to that of a buzzer or an ultrasonic transducer for determining distance in air: a piezoelectric ceramic disk bonded to a metal membrane and mounted in a housing. In a key finder, however, the housing is configured to form two Helmholtz resonators, each with its own resonance frequency. One of the resonators receives the signal and directs it to the flexing element. The circuit switches to transmit, and the flexing element emits a signal, via the second resonator, in response. After a sequence of sounds is transmitted, the circuit switches back to the receiving mode.

Figure 5.45 Key finder

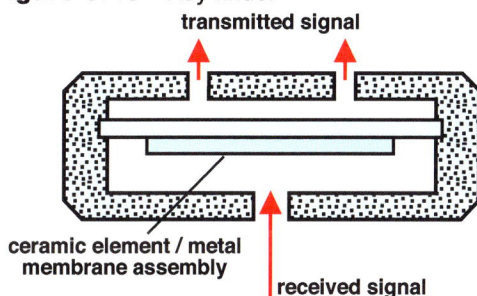

Transformers

Electromagnetic transformers are large and heavy and, at low outputs (30 W or less) they offer poor efficiency (approximately 30%). In contrast, piezoelectric transformers are small, thin, lightweight, and highly efficient (93-96% at outputs ≤ 30 W). Piezoelectric transformers offer additional significant advantages: they do not generate electromagnetic noise, which makes them ideal for use in computers, they deliver a more uniform output from a characteristically non-uniform input, and, because no insulation is used in their construction, they are non-flammable.

A piezoelectric ceramic transformer is composed of two ceramic components, an input component and an output component, typically combined in a single element. Two versions of the original piezoelectric transformer, the Rosen-type transformer, are shown in Figure 5.46. The input component of the device transforms the input electrical energy into mechanical (vibration) energy; the output component transforms the mechanical energy into output electrical energy. A piezoelectric transformer usually is operated at resonance frequency.

The voltage step-up ratio, r, (voltage out / voltage in) for a Rosen-type transformer can be determined from Equation 5.23:

Figure 5.46 Rosen-type piezoelectric transformers

(a) asymmetrical

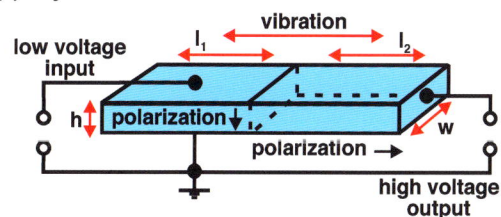

(b) symmetrical

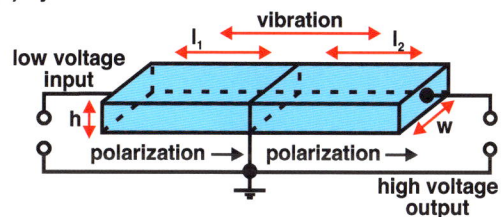

Figure 5.47 Electrical circuit equivalent for a (symmetrical) Rosen-type transformer

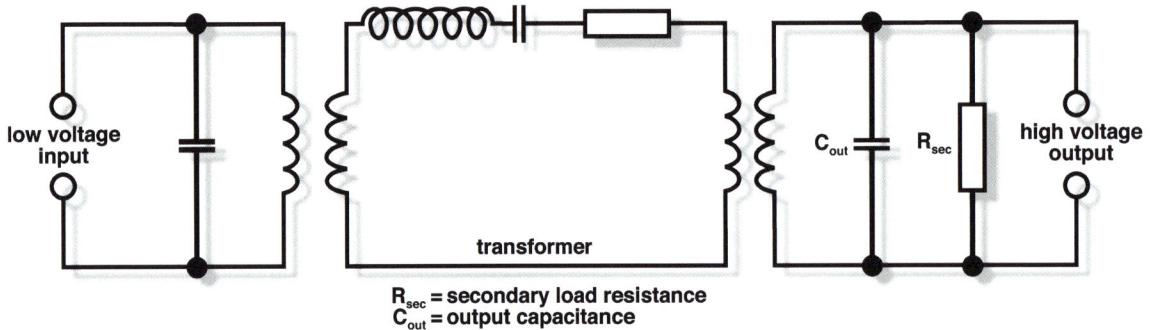

transformer

R_{sec} = secondary load resistance
C_{out} = output capacitance

Equation 5.23

$$r = (4 / \pi^2) (k_{31}k_{33}Q_m) (l_2 / h)$$
$$[2\sqrt{(s^E_{33} / s^E_{11})} / (1 + \sqrt{(s^D_{33} / s^E_{11})})]$$

where

k_{31}, k_{33} = electromechanical coupling factors
Q_m = mechanical quality factor for ceramic material
l_2 = output component length of ceramic element
h = height (thickness) of ceramic element
s^E_{33}, s^E_{11} = elastic compliance under constant electric field (short circuit)
s^D_{33} = elastic compliance under constant electric displacement (open circuit)

Alternatively, an approximate value for r can be obtained by using Equation 5.24:

Equation 5.24

$$r \cong k_1k_2Q_m$$

where

k_1 = electromechanical coupling factor for input component
k_2 = electromechanical coupling factor for output component
Q_m = mechanical quality factor for ceramic material

Equation 5.23 indicates that the step-up ratio can be increased by increasing the length of the output component, relative to the thickness of the ceramic element (l_2 / h). Equations 5.23 and 5.24 both indicate that r also can be increased by increasing the electromechanical

coupling factors (by using a different ceramic material, for example). However, because electrically loading the transformer reduces the mechanical quality factor for the ceramic material, r is load-dependent. Further, because the piezoelectric charge constants, d, are temperature-dependent, and electromechanical coupling factors, k, are directly correlated to d, the performance of a piezoelectric transformer can be affected by temperature changes.

Figure 5.47 shows an electrical circuit equivalent to a symmetrical Rosen-type transformer. Efficiency is maximum when the secondary load resistance, R_{sec}, and the output capacitance, C_{out}, are matched ($R_{sec} = 1 / \omega C_{out}$)*. However, because damping also is maximum when R_{sec} and C_{out} are matched, r is only $\sqrt{2}$, and output power is minimal. Damping values fall, and output power increases, as the relationship between R_{sec} and C_{out} deviates from optimal ($R_{sec} < 1 / \omega C_{out}$ or $R_{sec} > 1 / \omega C_{out}$).

For asymmetrical Rosen-type transformers, and for disks or other shapes, the geometry of the ceramic element also affects the performance of the transformer.

Although their performance qualities are, for the most part, excellent, Rosen-type piezoelectric ceramic transformers constructed in the original design are inherently mechanically weak, and in use they tend to break at the l_1 / l_2 interface. Modifications of the Rosen-type transformer, such as a planar ceramic element with a curved interface (Figure 5.48a) or a ceramic disk

* $\omega C_{out} = (2\pi)$ (resonance frequency, f_r) (C_{out}).

Figure 5.48 Modifications that improve the mechanical strength of a piezoelectric transformer

(a) Rosen-type transformer with curved interface

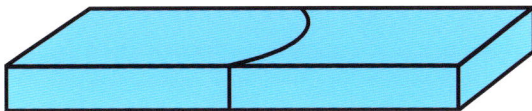

(b) disk-shaped element with curved interface

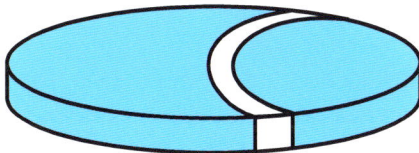

with a curved interface (Figure 5.48b) overcome this problem and enable the transformer to exhibit its potential.

The size, weight, and efficiency advantages of piezoelectric transformers, relative to electromagnetic transformers, and the absence of electromagnetic noise, are especially desirable in LCD backup inverters for laptop computers. A more uniform output, relative to a non-uniform input, makes these transformers very well suited for driving piezoelectric actuators in critical applications, including optics and medical equipment. Piezoelectric transformers also are used in fluorescent light tubes and x-ray tubes.

Composites

Composites are a diverse category of materials composed of a piezoelectric ceramic component admixed with a non-piezoelectric supporting material. A piezoceramic / rubber or piezoceramic / polymer composite consisting of ceramic particles embedded in a matrix of rubber or polymer, respectively, is called a 0-3 material, indicating that the ceramic particles are not in contact with one another, but that the supporting material is in contact with itself in three directions (Figure 5.49). The flexibility of the non-piezoelectric component allows the resulting material to be produced as highly flexible sheets or films, which are well suited for fabricating hydrophones (receivers). In the initial stages of their manufacture, 0-3 materials literally can be painted onto a surface, creating the potential for

thin layer sensory applications (e.g., vibration detection). Alternatively, the ceramic particles can be embedded in a rigid polymer.

Although they are difficult to prepare and to polarize, lead titanate materials are preferred for making 0-3 composites, because they exhibit stronger piezoelectric characteristics; lead zirconate titanates are most widely used in preparing other composites (6).

Composite materials in which the ceramic particles are in contact in one direction, such as fibers or thin rods of ceramic embedded in parallel in an inactive matrix, are called 1-3 materials (Figure 5.49). In 1-3 materials, the piezoelectric properties of the ceramic are employed in the thickness mode and, like 0-3 materials, they can be fabricated in large and flexible sheets, according to the characteristics of the supporting matrix. In comparison with 0-3 materials, 1-3 materials exhibit higher d_{33} values, a desirable characteristic for electromechanical actuators and acoustic transducers. Sensory capabilities of 0-3 and 1-3 materials are approximately equal.

Complex shapes, products with large surface area, and arrays can be constructed from 1-3 composites. Transducers for ultrasonic medical apparatus, nondestructive testing equipment for various materials, fish finders, and other devices are made by dividing a single ceramic plate into numerous small parts and embedding these parts in a polymer matrix. The transducer can be shaped to a specific design tailored to the demands of the application.

At the other end of the composite spectrum are 3-1 and 3-2 composites, which consist of a piezoelectric ceramic block with polymer-filled holes through one side (3-1), or both sides (3-2), perpendicular to the direction of polarization in the ceramic block. 3-1 and 3-2 composites are difficult to manufacture, show poor durability, and do not appear to have noteworthy properties, relative to more conventional materials.

Piezoelectric ceramic / polymer composites are particularly effective in underwater applications. The sensitivity of a piezoelectric element varies in proportion with the piezoelectric voltage constant, g, and the thickness of the element. For in-air applications, either the g_{33} (longitudinal mode) constant or the g_{15} (shear mode) constant of the element usually is put to use. In hydrostatic applications, however, sensitivity is proportional to constant g_h, which is equal to $g_{33} + 2g_{31}$ (6). For a single-piece piezoelectric ceramic element, $g_{33} \cong -2g_{31}$, thus the g_{33} and g_{31} contributions to g_h are subtractive, and the element exhibits poor sensitivity. Piezoelectric ceramic / polymer composites are designed to

Figure 5.49 Composite piezoelectric materials

(a) 0-3 material (piezorubber)

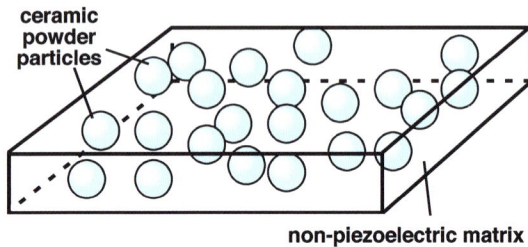

ceramic powder particles

non-piezoelectric matrix

(b) 1-3 material

ceramic elements

non-piezoelectric matrix

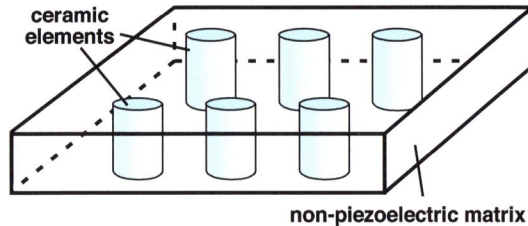

(c) 1-3 material (top view)

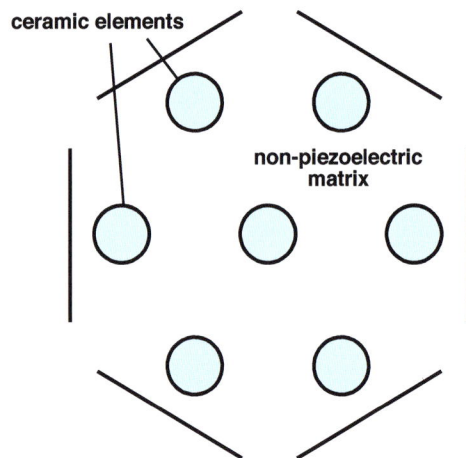

ceramic elements

non-piezoelectric matrix

eliminate either the g_{31} or the g_{33} contribution to g_h.

0-3 piezoelectric ceramic / polymer composites exhibit high sensitivity, high pressure tolerance, a broad operating bandwidth, and a good acoustical impedance match for air or water. 1-3 piezoelectric ceramic / polymer composites have approximately the same characteristics as 0-3 composites, but are lighter and more rigid, and are more easily tailored to specific applications. Piezoelectric ceramic / metal composites (moonies and other flextensional devices) are designed such that their g_{33} and g_{31} constants additively contribute to g_h, rather than subtractively contribute, and these materials exhibit very high sensitivity. On the other hand, ceramic / metal composites are more pressure dependant and have much narrower operating bandwidths than ceramic / polymer composites.

A piezoelectric composite can be an effective vibration damper. When the mechanical energy of vibration is transmitted to the composite it is converted into electrical energy by the piezoelectric component of the composite. The electrical energy is, in turn, converted to Joule heat by a resistor connected in series with the composite. Thus, energy is removed from the system by the resistor (otherwise the electrical energy would be efficiently reconverted to mechanical energy), and the vibration is damped. Damping is most rapid when the resistance, R, $= 1 / (2\pi f C)$, where f is the frequency of the vibrations to be damped and C is the capacitance of the piezoelectric (composite) material. Including a small amount of carbon black in the composite (6-7 volume percent) significantly changes the electrical conductivity of the composite and enables it to dissipate vibrational energy without connecting it to a resistor.

References

1. Uchino, K., *Ferroelectric Devices*
   Marcel Dekker, New York (2000).
2. Hughes, W.J., *Transducers, Underwater Acoustic*
   Vol. III/Appendix 30 in *Acoustic Transduction
   — Materials and Devices*
   Annual Report (1 January 1998 to 31
   December 1998) Office of Naval Research,
   Contract No: N00014-96-1-1173 (April 1999)
3. Wang, H., T. Ritter, W. Cao, and K.K.
   Shung, *Passive Materials for High Frequency
   Ultrasonic Transducers*
   SPIE Conference on Ultrasonic Transducer
   Engineering, San Diego, CA, Feb. 1999,
   SPIE Volume 3664, 35-42 (1999).
4. Alkoy, S., A.C. Hladky, A. Dogan, J.K.
   Cochran, Jr., and R.E. Newnham, *Piezoelectric
   Hollow Spheres for Microprobe Hydrophones*
   Vol. III/Appendix 43 in *Acoustic Transduction
   — Materials and Devices*
   Annual Report (1 January 1999 to 31
   December 1999) Office of Naval Research,
   Contract No: N00014-96-1-1173 (June 2000)
5. Roveti, D., *Ultrasound Power Measurement*
   Medical Electronics, 98-106, Dec. 1989.
6. Tressler, J.F. and K. Uchino, *Piezoelectric
   Composite Sensors*
   Vol. II/Appendix 37 in *Acoustic Transduction
   — Materials and Devices*
   Annual Report (1 January 1999 to 31
   December 1999) Office of Naval Research,
   Contract No: N00014-96-1-1173 (June 2000)

Request annual reports to Office of Naval
Research from:
 Office of Naval Research
 Regional Office Chicago
 536 S. Clark Street
 Room 208
 Chicago, IL 60605-1588.

miscellaneous

This chapter summarizes important peripheral information about using piezoelectric ceramic elements, specifically: securing an element, making electrical connections, and testing the performance of the element.

Securing a Piezoelectric Element

Of the three approaches to securing an element—gluing, soldering, and clamping—the best approach usually is gluing. An epoxy or acrylate glue will provide a strong, flexible, non-conducting bond.

Soldering is a less reliable means of securing a ceramic element because, over time, vibrations in the element could cause the bond to fail. Also, the high temperatures associated with soldering must be applied carefully, to avoid damaging the piezoelectric properties of the element. On the other hand, there is no alternative to soldering if an electrically conductive connection is needed.

Clamping is the least reliable and most restrictive means of securing an element, but is fast and simple, and avoids subjecting the ceramic to potentially damaging elevated temperatures.

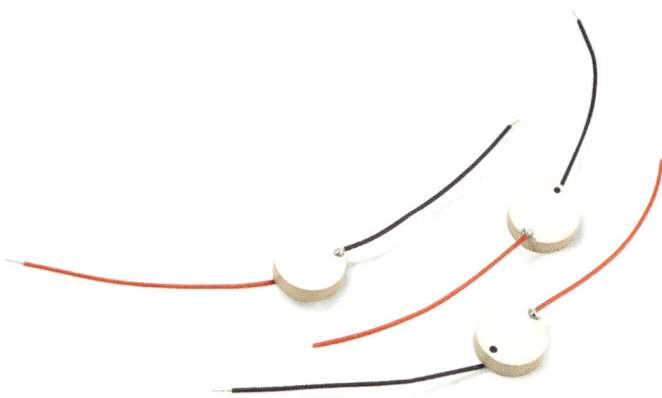

Electrical Connections

The general approach described in this chapter is widely useful for soldering electrical connections to a piezoelectric element.

Performance Testing

The piezoelectric charge constant, d, is the mechanical strain experienced by a piezoelectric material per unit of electric field applied, or, alternatively, is the polarization generated by the material per unit of mechanical stress applied. Defects in a ceramic caused by errors in formulating, processing, polarizing, or subsequently handling and storing a ceramic element are reflected by a low d_{33} value. Consequently, a fast, easy means of determining d_{33} values is invaluable for characterizing a piezoelectric ceramic, to ensure quality and unit to unit consistency. Modern d_{33} testing units meet these performance needs.

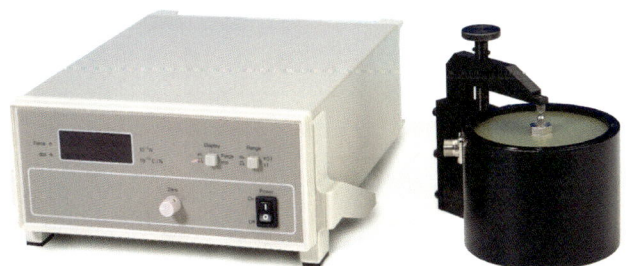

Miscellaneous

Securing a Piezoelectric Ceramic Element

In order to accomplish its task, a piezoelectric ceramic element usually must be secured to a solid surface. Of the three approaches—gluing, soldering, and clamping—the overall best approach usually is gluing. An epoxy or acrylate glue will provide a strong, flexible, non-conducting bond. The flexible nature of a glue bond usually eliminates the possibility of vibration-induced fatigue during long-term operation. When needed, glues are available that can be used at temperatures exceeding 150°C. Such temperatures will encompass the recommended operating limit for most piezoelectric ceramic materials. Be aware, however, that high temperature programs needed to set some glues could affect the piezoelectric characteristics of the ceramic element. Compare the Curie point and recommended maximum operating temperature of the ceramic with the curing temperature of the glue; if there is overlap, contact the manufacturer about the feasibility of prolonging the curing time at a lower temperature—or use another glue.

Soldering is a less reliable means of securing a ceramic element because, over time, vibrations in the ceramic can cause the bond to fail. Also, the high temperatures associated with soldering must be applied carefully to a ceramic element—the Curie point of a polarized ceramic can be well below the temperature of hot solder. On the other

Table 6.1 General soldering procedure

Equipment and Materials
Soldering iron (~350°C)
Solder (SN 96: 96% tin / 4% silver, 0.032" diameter)
Flux (active rosin) (Kestor 1544 or equivalent—no ZnCl or other corrosive agents)
Leads (common sizes 28AWG-32AWG)
Small, sharp blade (X-acto® knife or equivalent)
Solvent (for removing excess flux)
Cotton swabs

Procedure
1. Turn on the soldering iron and allow it to stabilize at ~350°C.
2. Using X-acto knife, remove conductive coating (if present) from areas where leads are to be attached.
3. Wipe prepared surface with solvent and allow to dry.
4. Flux and tin wire with solder.
5. Place small amount of flux on wire and solder area.
6. Melt small amount of solder onto soldering iron tip.
7. Place wire on surface in desired position.
8. Place iron with solder on wire and surface. Hold for 1-2 seconds, then remove iron. Solder should flow from iron to wire and surface.
9. Allow solder joint to cool before moving wire. Repeat steps 2-8 for each connection.
10. Clean solder joint(s) with solvent.

hand, a soldered connection is electrically conductive and, in some situations, this characteristic might be needed.

Table 6.1 describes a generalized soldering procedure that can be adapted for securing a ceramic element into place. Before soldering a ceramic element be sure to read carefully the manufacturer's recommendations for the type of solder to be used, pre-soldering treatment of the surfaces to be soldered, etc. To minimize the effects of heat on the piezoelectric properties of the ceramic, always keep the soldering time as short as possible (3 seconds or less is ideal).

Remember that piezoelectric ceramic elements have substantial pyroelectric coefficients. The thermal energy introduced by soldering an element into place—or by soldering an electrode to an element—will generate an electrical charge in the element. To avoid the unpleasant consequences associated with discharging the element, apply solder to a piezoelectric ceramic element with the element in short-circuit conditions.

The least reliable means of securing an element is clamping. In addition to the potential for damaging the element while securing it, vibrations in the working element can enable it to slip out of the clamp. Clamping also is the most restrictive approach, minimizing the freedom of movement that can be essential to the best performance by the ceramic element. On the positive side, clamping is fast and simple, and avoids subjecting the ceramic to potentially damaging elevated temperatures.

Electrical Connections

Many ceramic elements from APC International, Ltd. are available either with or without electrical leads. If the stock leads on a product do not fit your needs, we will be happy to discuss custom-manufacturing the item to meet your unique specifications.

When soldering electrical connections to an APC piezoelectric ceramic element, follow the recommendations in Table 6.1 (type of solder and flux to be used, etc.). To minimize the effects of heat on the piezoelectric properties of the ceramic, always keep the soldering time as short as possible (3 seconds or less is ideal).

The thermal energy introduced by soldering an electrode to an element will generate an electrical charge in the element. To avoid the unpleasantness that accompanies discharging a charged piezoelectric ceramic element, apply solder to an element only under short-circuit conditions.

Testing

The piezoelectric charge constant, d, is, alternatively, the mechanical strain experienced by a piezoelectric material per unit of electric field applied, or the polarization generated by the material per unit of mechanical stress applied. Defects in a ceramic caused by incorrect composition, incorrect processing, improper handling, insufficient polarization voltage or time, etc., are reflected by a low d_{33} value.

A d_{33} testing unit is a fast, reliable, low cost means of determining d_{33} values and thus is invaluable for characterizing a piezoelectric ceramic, to ensure quality and unit to unit consistency. Operation simply involves securing the piezoelectric element into the tester and reading the d_{33} value from a display. The most versatile models accommodate single crystals and polymers as well as ceramic elements of varying size and shape, and allow simultaneous testing of multiple elements.

Figure 6.1 Piezoelectric charge constant (d_{33}) testers

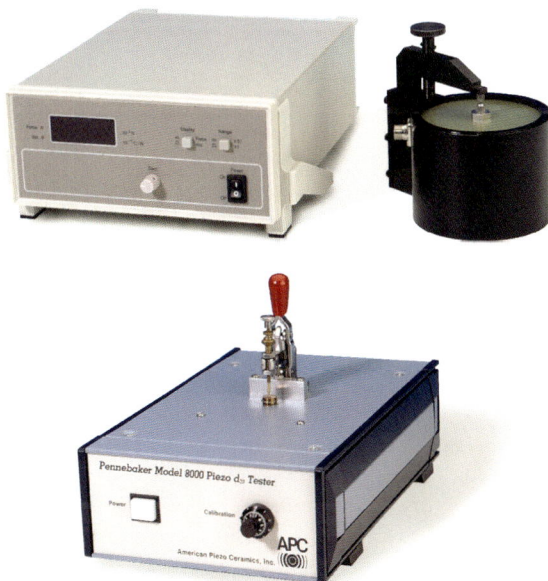

APC International, Ltd.

APC International, Ltd.
213 Duck Run Road
P.O. Box 180
Mackeyville, PA 17750 USA

P: +1.570.726.6961
F: +1.570.726.7466
sales@americanpiezo.com
www.americanpiezo.com

APC
International, Ltd.

CPSIA information can be obtained
at www.ICGtesting.com
Printed in the USA
LVIC04n2030231013
358276LV00016B/145